6/20/06

For Patti Suleima

Thanks for a
inspirational dialogue

Peter Hall

SWARM CREATIVITY

Swarm Creativity

Competitive Advantage through
Collaborative Innovation Networks

Peter A. Gloor

OXFORD
UNIVERSITY PRESS

2006

OXFORD
UNIVERSITY PRESS

Oxford University Press, Inc., publishes works that further
Oxford University's objective of excellence
in research, scholarship, and education.

Oxford   New York
Auckland   Cape Town   Dar es Salaam   Hong Kong   Karachi
Kuala Lumpur   Madrid   Melbourne   Mexico City   Nairobi
New Delhi   Shanghai   Taipei   Toronto

With offices in
Argentina   Austria   Brazil   Chile   Czech Republic   France   Greece
Guatemala   Hungary   Italy   Japan   Poland   Portugal   Singapore
South Korea   Switzerland   Thailand   Turkey   Ukraine   Vietnam

Copyright © 2006 by Peter A. Gloor

Published by Oxford University Press, Inc.
198 Madison Avenue, New York, New York 10016

www.oup.com

Oxford is a registered trademark of Oxford University Press

Library of Congress Cataloging-in-Publication Data
Gloor, Peter A. (Peter Andreas), 1961–
Swarm creativity : competitive advantage through collaborative
innovation networks / Peter A. Gloor.
p.   cm.
Includes bibliographical references and index.
ISBN-13: 978-0-19-530412-1
ISBN 0-19-530412-8
1. Business networks. 2. Information networks. 3. Group decision making.
4. Teams in the workplace. 5. Creative ability in business. 6. Technological
innovations—Management. 7. Knowledge management.   I. Title.
HD69.S8.G586 3006
338.8'7—dc22   2005011977

9 8 7 6 5 4 3 2 1

Printed in the United States of America
on acid-free paper

ACKNOWLEDGMENTS

It is fitting that a book about swarm creativity and collaborative innovation networks (COINs) should result from a COIN that itself worked in swarm creativity. This book is not the work of a single person, but resulted from the interactions of a large and wonderful group of people. Without the generous help of this group of friends, it would never have been completed.

Several individuals deserve mention. My first thanks go to Thomas Malone, my generous host for the last three years at the Massachusetts Institute of Technology (MIT) Center for Coordination Science (CCS), and to Thomas Allen of the MIT Sloan School of Management. Both have been most inspiring mentors. I am also grateful to Rob Laubacher, John Quimby, George Hermann, and Peggy Nagel, all colleagues at CCS, for many thought-provoking discussions.

Yan Zhao developed countless versions of the temporal communication flow analysis (TeCFlow) software tool, putting up with myriad requests for change and always adding her own excellent ideas and insights. Our work at Dartmouth College has been generously supported by Hans Brechbuhl and M. Eric Johnson at the Tuck School of Business Center for Digital Strategies, and by Fillia Makedon from the Dartmouth Devlab.

My former colleagues at Deloitte have been instrumental in developing the original concept of collaborative knowledge networks (CKNs) and COINs. I am deeply indebted to Robin Athey, cofounder of the first COIN on CKNs. Elmar Artho, Jessica Bier, Emma Connolly, Niki Flandorfer, Anne Gauton, Laura Ghezzi, Karin Guentensperger, Adriaan Jooste, Roland Haenni, Marc

Killian, Nico Kleyn, Steven Lambert, Thomas Ojanga, Thomas Ritzi, Stuart Rosenberg, Martin Spedding, Andreas Timpert, Gilbert Toppin, Virginia Villalba, and—most of all—Thomas Schmalberger were crucial contributors to this COIN. I am also truly grateful to my colleagues in the earliest COINs. Brian Dunkel, Scott Dynes, Irene Lee, and Angel Velez-Sosa in the Animated Algorithms project, Rade Adamov, Brian Dunkel, Norbert Hoffmann, Peter Huegli, Wolfgang Luef, Zoltan Majdik, Rudolf Marty, and Kurt Wolf at UBS, and Stephan Carsten, Emanuele Fossati, Sandra Giacalone, Hans Peter Hochradl, Marc Killian, Till Knorr, Rudolf Lehmann, Franz Odekerken, Christoph Sand, Roland Stadler, Ragnar Wachter, and Jill Williams at Deloitte were influential in developing my early understanding of the workings of COINs.

My friends and colleagues Hermann Blaser, Scott Dynes, Bill Ives, Takis Metaxas, Andre Ruedi, Suzanne Spencer-Wood, and Wayne Yuhasz helped me through the ups and downs of the gestation process, reading and commenting on various versions of the manuscript, offering great advice, and contributing their own excellent ideas.

I am grateful to the late David Hoenigsberg, Douglas Grannell, and Gareth Dylan Smith; to Fritz Bircher and Reinhold Krause; to Christoph Von Arb and Pascal Marmier; and to Richard E. Deutsch, Martin Duerst, Erich Gamma, Stefano Mazzocchi, and David Tennenhouse for sharing the insights they gained in leading their own COINs.

Cathy Benko, Rob Cross, Doug Downing, Walter Etter, Jan Fuelscher, Dirk Havighorst, Carey Heckmann, Bruce Hoppe, Marcel Isenschmid, Charles Leiserson, Bernardo Lindemann, Stefano Pelagatti, Gregor Schrott, Gary Shepherd, Marshall Van Alstyne, and Daniel Wild all offered excellent input in their reviews of earlier versions of this manuscript.

I am deeply indebted to Scott Cooper for adding his own insights while making my writing much more accessible. I am also grateful to John Rauschenberg at Oxford University Press for his support in publishing this book.

This book is dedicated to my children, Sarah and David, my best teachers about what matters.

CONTENTS

SWARM CREATIVITY

INTRODUCTION

At the Tipping Point

> As managers, we need to shift our thinking from command
> and control to coordinate and cultivate—the best way to gain
> power is sometimes to give it away.
> —*Thomas W. Malone, The Future of Work* (2004)

We are at the dawn of a new way of working together, thanks in large part to technological advances that allow a more radically superior mode of innovation than ever before. Soon, businesses throughout the world will be looking for ways to unleash the power of their COINs. Some will be scrambling to figure out how to innovate in the new business environment.

Few today know COINs—*collaborative innovation networks*—by that name, even though COINs have been around for hundreds of years. Many of us have already collaborated in COINs without even knowing it.

What makes them so relevant today is they have reached their *tipping point.*

This idea of a tipping point comes from Malcolm Gladwell, who describes the moment of critical mass where radical change is more than just possible, but is a certainty. Thanks to the communication capabilities of the Internet, COINs are at that threshold; they have achieved global reach and they can spread at viral speed. Gladwell uses the word "epidemic" to describe what happens at the tipping point.[1]

What is a COIN? There is a lot to the answer, and this book delves deeply into the DNA of COINs—the traits that must exist for them to exist and for them to succeed. Here, we offer a brief definition:

> A COIN is a cyberteam of self-motivated people with a collective vision, enabled by the Web to collaborate in achieving a common goal by sharing ideas, information, and work.

In a COIN, knowledge workers collaborate and share in internal transparency. They communicate directly rather than through hierarchies. And they innovate and work toward common goals in self-organization instead of being ordered to do so. Working this way is key to successful innovation, and it is no exaggeration to state that *COINs are the most productive engines of innovation ever.* COINs have produced some of the most revolutionary drivers of change of the Internet age, such as the World Wide Web and Linux.

This book explains COINs in depth, makes the case for why businesses ought to be rushing to uncover their COINs and nurture them, and provides tools for building organizations that are more creative, productive, and efficient by applying principles of creative collaboration, knowledge sharing, and social networking. It contains information on leveraging COINs to develop successful products in research and development, grow better customer relationships, establish better project management processes, and build higher-performing teams. It even offers a method for locating, analyzing, and measuring the impact of COINs on an organization.

In short, this book answers four key questions:

1. *Why* are COINs better at innovation than conventional organizations?
2. *What* are the key elements of COINs?
3. *Who* are the people that participate in COINs, and how do they become COIN members?
4. *How* does an organization transform itself into a COIN?

In subsequent chapters, you will learn about the *creators* who come up with the visionary ideas, the *communicators* who serve as ambassadors of COINs and help carry new inventions over their tipping points, and the *collaborators* who form the "glue" of a COIN and make it see the vision through to reality.

In a sense, this is really two books in one. Chapters 1 through 6 make the business case for COINs and explain the technological advances that have led to their tipping point. Appendixes A through C are tools and steps to COINs: how to uncover COINs by analyzing the evolution of communication patterns; how to leverage COINs and combine customers and suppliers into a seamlessly integrated value network; and how to reassign tasks of COINs. One of these tools also shows individuals how to become efficient members of COINs and fully leverage their own skills as creators, communicators, and collaborators.

The book is backed by numerous case studies of different types. Principles are illustrated by famous COINs, such as the groups that created the Web and Linux, as well as by stories of how "ordinary people" are leveraging COINs to achieve extraordinary success. The high-profile successes illustrate the tremendous potential of COINs, and the "everyday" COINs demonstrate that anyone can initiate and succeed with a COIN. Chapter 5 presents some specific examples from a leading manufacturer, a global service provider, financial institutions, and other sectors.

Innovate—Collaborate—Communicate

The creation of the World Wide Web is such an exemplary COIN that it warrants some further mention in this introduction. When Tim Berners-Lee introduced the World Wide Web to the academic world at the 1991 Association for Computing Machinery (ACM) Hypertext Conference in San Antonio, Texas, the hypertext concept had already come a long way.[2]

A series of visionaries before Berners-Lee proposed the basic ideas behind the Web, namely to link pieces of information and make them accessible to many users. The first was Vannevar Bush, the famous scientist and advisor to Franklin Delano Roosevelt. In a 1945 magazine article, he described a system called Memex to make and follow links.[3] But the information technology that emerged in the 1950s consisted of microfiches and card readers, and Bush's Memex system was not built in his lifetime. In the 1960s, other visionaries moved the idea forward. Ted Nelson, who coined the term "hypertext," and computer scientist Douglas Engelbart, who demonstrated the first working hypertext system, took up Bush's idea. Engelbart invented the computer mouse while building the prototype "oNLine System" (NLS) that provided hypertext browsing, editing, and e-mail. However, both men were still ahead of their time, and the spark did not fully ignite.

In the late 1980s, the hypertext concept found a firm footing in the academic computer science community, bringing together hundreds of researchers at the annual hypertext conferences. But it was only in the early 1990s that technology and society were ready for hypertext, and a team of enthusiasts collaborated to spin the Web.

Tim Berners-Lee wrote his first hypertext system—Enquire—in 1980, while working as a consultant for CERN, the European Organization for Nuclear Research laboratory. It took him years of grassroots lobbying at CERN, writing and circulating research proposals, until his boss finally approved the purchase of a NeXT computer in 1990s and allowed Berners-Lee to go ahead and write a global hypertext system. During the same year, Robert Cailliau, another hypertext enthusiast from the CERN computer group, joined the effort. Together, they developed the first Web browser, editor, and server, soon reinforced by a small team of volunteers, mostly summer students at CERN.

Finally, in December 1991 at the San Antonio conference, Berners-Lee and Cailliau presented the ideas of their group to the academic community at large. By that time, the fervor had already spread, and the first Web server outside Europe had been installed. People flocked to the Web development

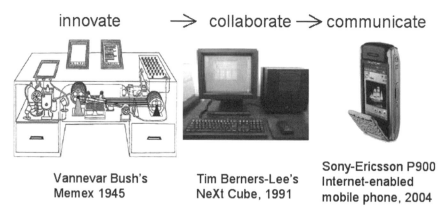

innovate $\rightarrow$ **collaborate** $\rightarrow$ **communicate**

| Vannevar Bush's Memex 1945 | Tim Berners-Lee's NeXt Cube, 1991 | Sony-Ericsson P900 Internet-enabled mobile phone, 2004 |

Figure 1.1.   From the Memex to the ubiquitous Internet. Memex photo from Bush, "As We May Think" (1945); NeXt cube photo courtesy of NeXt, Inc.; Sony Ericsson P900 photo courtesy of Sony Ericsson.

team, and programmers from Finland, Austria, Germany, France, and California built new versions of Web browsers and servers. In February 1993, when Marc Andreessen of the National Center for Supercomputing Applications (NCSA) at the University of Illinois released a browser called "Mosaic," the tipping point was reached. The Web exploded. The following year, Andreesen and his colleagues left NCSA to form Mosaic Communications Corporation, which later became Netscape and turned the twenty-somethings into very young millionaires.

The birth and explosive growth of the Web exhibit all the characteristics of a highly successful COIN at work. In the next chapters, we'll explore the "swarm creativity" that resulted in the Web, learn how internal transparency creates efficiency and leads to success, and examine how virtual trust and an ethical code hold a COIN together so it can fulfill its vision.

To make the business case, though, we begin with benefits in chapter 1. COINs offer tremendous innovative power, and we'll see that Thomas Malone's words from the beginning of this introduction are true. If working collaboratively, in a transparent environment, is "giving away" power, it is also the way to gain the power of COINs.

CHAPTER 1

COINs and Their Benefits

Late in 1999, I was riding the train in Switzerland from Basel to Zurich when my mobile phone rang. A colleague of mine wanted to know if I could be on a conference call, right away, with some people from DaimlerChrysler. They had a problem.

Gary Valade, DaimlerChrysler's executive vice president and head of its Global Procurement and Supply division, was facing a big challenge. He wanted to capitalize on emerging technologies and streamline his function, optimize transaction processes, tighten control of materials flow and inventory, and improve the transfer of information throughout the company's supply chain. He needed to do this in order to address the competitive pressure from other players in the automotive industry; DaimlerChrysler's competitors were also busy leveraging e-business technologies in automotive procurement and supply chain processes. Whatever solution was developed had to be implemented on a worldwide basis.

DaimlerChrysler ultimately solved the problem and answered the challenge with its e-extended Enterprise (e^3) initiative. This initiative has enhanced DaimlerChrysler's supplier relationships, increased efficiencies, reduced new product development time, and cut expenses. We'll learn more about the project in chapter 5. What is important here is *how* the project unfolded.

The e^3 project began in January 2000. At DaimlerChrysler, some 15 people from the Global Procurement and Supply function worked full-time, supported by members of the information technology (IT) organization and by part-time help from employees of the business units, as required. A team of 15 external consultants from the United States and Europe reinforced the project team. This effort led to a successful outcome: the company spent slightly more than $10 million to optimize its procurement function, and it saved more than $50 million in the first year of operation.

Although it wasn't planned from the beginning as such, the project team operated as a global collaborative innovation network, or COIN.

DaimlerChrysler met a tremendous challenge successfully, thanks in large part to a new way of working together in a business environment driven more by knowledge than ever before. Today, collaborating in an open and transparent flow of knowledge is key to successful innovation, and COINs are the most productive engines of innovation ever. COINs have produced some of the most revolutionary drivers of change in the Internet age. One example is Linux. Another is the World Wide Web itself.

What makes COINs better? They allow for building organizations that are more creative, productive, and efficient by applying principles of creative collaboration, knowledge sharing, and social networking. Sponsors and members of COINs often change their work and leadership styles to become more creative innovators, more efficient communicators, and more productive collaborators. The evidence shows that COINs can be leveraged to develop successful products in research and development (R&D), grow better customer relationships, establish better project management processes, and build high-performing teams. COIN-enabled organizations demonstrate more efficient leadership, culture, structure, and business processes.

How did it happen that the team from Switzerland—a small landlocked country—won the 2003 America's Cup, sailing's most coveted prize? The answer illustrates the three words that make up the COIN acronym. Alinghi, the Swiss sailing team, functioned as a close-knit, *collaborative* community of more than 100 incredibly motivated experts and athletes from 15 countries. Ernesto Bertarelli, the Swiss biotechnology billionaire who provided the financial backing for the team, was just another team member. Alinghi developed superior technology by creatively applying research insights from other areas; consider, for instance, the team's *innovation* in fluid dynamics, which came through collaboration with a Swiss university medical researcher who was modeling the flow of blood vessels in the human body. Today, Alinghi continues to generate and test its innovative ideas through a research *network* of five universities. Dirk Kramers, Alinghi's chief designer, is very clear in his praise—not for geniuses working in isolation, but for collaborative networks of highly motivated researchers. Team Alinghi illustrates how organizations operating as COINs become leaders of innovation, flexibility, and efficiency.

COIN applications range from IT outsourcing, sales force optimization, R&D, mergers and acquisitions, software, and distributed product development to running political campaigns and charities online. Projects in companies such as DaimlerChrysler, Union Bank of Switzerland (UBS), Novartis, Intel, Deloitte, HP, and IBM have successfully operated as COINs to deliver innovations more rapidly and at lower costs. You'll learn about some of those projects in later chapters.

What Makes a COIN?

Members of COINs self-organize as *cyberteams,* teams that connect people through the Internet—enabling them to work together more easily by communicating not through hierarchies, but directly with each other. The individuals in COINs are highly motivated, working together toward a common goal—not because of orders from their superiors (although they may be brought together in that way), but because they share the same goal and are convinced of their common cause.

People in COINs usually assemble around a new idea outside of organizational boundaries and across conventional hierarchies. They innovate together in *swarm creativity*—which you'll learn about in the next chapter. By "swarming together," they work together in a structure that enables a fluid creation and exchange of ideas. Looked at from the outside, the structure of a COIN may appear chaotic, like a bee or ant colony, but it is immensely productive because each team member knows intuitively what he or she needs to do.

COIN members develop new ideas as a team; the whole is much more than the sum of its parts. They create and share knowledge in an environment of high trust by collaborating under a common code of ethics.

In short, three characteristics define COINs. They:

1. *Innovate* through massive collaborative creativity
2. *Collaborate* under a strict ethical code
3. *Communicate* in direct-contact networks

COINs Deliver Benefits to Organizations

The benefits to organizations that embrace the concept of COINs are substantial.

COINs make organizations more innovative and more collaborative. Enabling and investing in COINs is an enormously cost-efficient way to innovate and get ahead of the competition, as DaimlerChrysler's experience shows. How? Organizations that make supporting COINs an integral part of their culture become more innovative by creating a transparent innovation process and making it possible to reward hidden innovators. People in organizations that support COINs become more efficient, as well as effective at working together. Tim Berners-Lee was working at CERN, the European nuclear physics consortium, when he began to create the World Wide Web. Although the effort was far afield of CERN's core competency, the organization let him proceed and won considerable recognition for doing so. A company can always decide later whether innovations fit into the strategy and whether to invest in them.

COINs make organizations more agile. Organizations that support COINS are better equipped to react to market and technology changes. DaimlerChrysler showed that by leading the way in establishing the main online car parts supplier marketplace. And UBS became one of the first banks to have a Web-based intranet—at very low cost. (This is detailed in chapter 5.)

COINs infuse external knowledge into the organization. COIN members have knowledge of what is going on in the marketplace, and they usually take initiative in building up personal links with peers outside the organization. By supporting COINs, organizations can improve and build on that knowledge. The core COIN team at DaimlerChrysler was well aware of what General Motors and Ford were up to, and was able to advise its own management to join forces with those companies to establish an online car parts marketplace. At UBS, knowledge of the ease of use and simplicity of Web technologies convinced the team to ignore the initial resistance and move forward with the "Bank Wide Web."

COINs uncover hidden business opportunities. Supporting COINs allows a company to identify hidden business opportunities well before the competition. As previously mentioned, at DaimlerChrysler, a COIN allowed the company to participate in the largest online car parts marketplace before its European and Asian competitors. At IBM, a small COIN of midlevel managers was first to spot the significance of the Internet for the company's business. Lobbying tirelessly, the COIN members convinced top management to put IBM's full weight behind e-business—which generated $40 billion in revenue for the company in 1999.[1]

COINs release synergies. At near-zero cost, Deloitte Consulting in Europe was able to leverage existing resources—both people and infrastructure—as a COIN to create an entirely new service offering and generate lots of new revenue. (This is detailed in chapter 5.)

COINs reduce costs and cut time to market. By supporting nimble and flexible COINs, organizations create a more transparent environment, which leads to greater trust and, in turn, to reduced transaction costs. Time-to-market for new inventions can be reduced substantially. In fact, COINs offer an enormously cost-efficient way to invest in innovations. People normally join a COIN not for immediate monetary reward, but because they are fascinated by a problem and want to contribute to solving it. This was true for the Deloitte Consulting practice, which not only was set up in a few months at no cost, but also—at the height of the dot.com craze—served as a highly efficient and low-cost retention device for highly marketable staff.

COINs help organizations locate their experts and reward hidden contributors. By supporting COINs, an organization can identify its internal sources of tacit knowledge and innovation. Frequently, these people are not high up in the official hierarchy and are not recognized by management. If supported appropriately, though, they offer tremendous value. By recognizing that value and making the merits of these people more obvious, an organization not only improves morale but also enhances organizational productivity and profitability. At DaimlerChrysler, the tremendous contributions by a group of people from outside the "official" project became obvious, thanks to the COIN structure of the project. Once on top management's radar screen, these people were quickly promoted to official leadership positions within the project and the business units that grew out of the project. At UBS, it became obvious that the skills and tutoring of one particular programming wizard had been crucial to the success of the project. He became one of the first non-managers named a vice president at the bank.

COINs lead to more secure organizations. The higher transparency and the high-trust operating environment of COINs have significant benefits, as transparent communication flow exposes security risks early on. This is apparent in the examples of several COINs, such as the Internet Engineering Task Force (IETF), the World Wide Web Consortium (W3C), and Linux developers. They operate in complete transparency: everybody posting to a mailing list in IETF, W3C, and Linux working groups knows that everyone else can read their postings. Impostors, spammers, and anyone trying to use the groups and their lists for any reason other than to benefit the community are immediately exposed, policed, and—if unwilling to rectify their behavior—expelled.

These are some of the many quantitative and qualitative benefits organizations can realize through COINs. We'll explore them more fully in subsequent chapters. But why should individuals be willing to dedicate their time and collaborate in COINs? If organizations aren't encouraging their people to work in COINs, is it worth it? What happens if senior management doesn't see the value?

COINs Benefit Individuals

Even without initial managerial backing, and in the face of organizational obstacles, there are still strong incentives for individuals to join fledgling COINs.

People initially join COINs because they are fascinated by the challenge and care deeply about the goals of their community. Their primary currency of reward is peer recognition. Programmers participate in open-source software projects such as Linux because they want to further the development of their product, and because they want to build a reputation as Web or Linux experts. Consultants joined Deloitte's e.Xpert group because they wanted to become e-business experts ("e.Xperts") and be recognized as such. DaimlerChrysler's project was started because team members wanted to become recognized as e-procurement experts. The Bank Wide Web at UBS was started because the core team believed in the future of the Web and wanted to apply it to daily business tasks.

There are more tangible benefits for individuals, as well.

COINs help individuals build a wider network that bridges "structural holes" and get direct access to business-critical information. COIN members get to know deeply committed people with shared values and beliefs whom they might not otherwise have met. This allows COIN members to build up an invaluable network of people to draw on in professional situations—and fill in the "structural holes" in their existing personal networks. The members of the Deloitte Consulting e.Xpert COIN tapped into the skill base of their fellow e.Xperts to complete successfully their own e-business engagements. Further, people participating in COINs discover links and synergies they never knew existed. Members of the e.Xpert network at Deloitte Consulting were tipping each other off on new business leads and e-business project opportunities well before requests came in through official channels. This meant that their utilization—one of the primary measures of success for a consultant—on average was substantially higher than that of the rest of the Deloitte consultants.

COIN members build personal relationships with leaders in their field. People participating in COINs learn about experts in specific problem do-mains. They have better opportunities to leverage the skills and knowledge of others—far beyond those normally advertised in the corporate yellow-page directories. For example, the COIN members in DaimlerChrysler got to know luminaries of Internet commerce from leading software vendors such as SAP, CommerceOne, Manugistics, i2, and Ariba. This means that in their future activities—even outside of the specific project—individual COIN members can call on some of the world's most highly valued experts in leading-edge technologies.

COIN members learn new skills, become experts, and often find themselves promoted—rapidly. Over time, people participating in COINs themselves become experts. They become more flexible in adapting to market and technology changes; they obtain access to innovations that are under way in their organization. This creates new opportunities for personal development and improves prospects for personal advancement. For example, people working on the Bank Wide Web project at UBS became early Webmasters and Internet experts. They had an edge well before the dot.com hype, and later found themselves promoted rapidly within the bank or lured to high-paying positions with outside consulting firms.

Transforming Organizations to Support COINs

The blood system energizes the human body, but the human body also needs the nervous system to steer and supervise its vital functions. Similarly, while a company's lifeblood flows through its business processes, the long-term health of the company depends in large part on its knowledge flow (figure 1.1).

Over the past two decades, businesses have largely focused on streamlining structured business processes. Today, the challenge is to optimize the flow of knowledge, streamlining unstructured, knowledge-intensive innovation processes, and turning organizations into COINs. By visualizing the flow

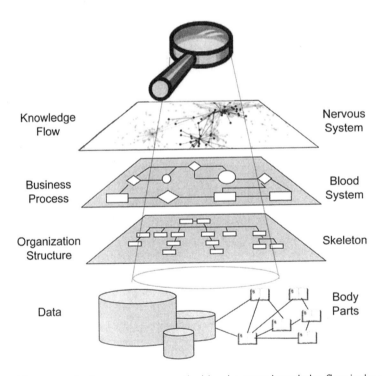

Figure 1.1.   Business processes are the blood system; knowledge flow is the nervous system.

of knowledge, making it transparent, and optimizing its course, organizations and individuals become more creative, innovative, and responsive to change. This is one of the keys to success in the new century. But while the importance of continuously optimizing and fine-tuning business processes is universally recognized, the importance of redesigning and optimizing knowledge flows is still widely underrecognized.

Organizations can successfully promote COINs by giving up central control in favor of self-organization in swarm creativity, developing an ethical code, and setting up a social network connected by hubs of trust (which you'll also learn about in subsequent chapters). The lessons learned for COINs and cyberteams apply to every company and organization in daily business life. Companies that have already harnessed the principles of COINs have benefited by successfully creating new, innovative products or making existing processes more flexible, efficient, and agile.

COINs will be the foundation of virtual teamworking for tomorrow's increasingly virtual global enterprises. Innovation is crucial to the long-term success of an organization, and COINs are the best engines to drive innovation. To understand how and why, and to see how results from the companies presented in this book relate to your company, you need to understand swarm creativity. That is the subject of the next chapter.

CHAPTER 2

Collaborative Innovation through Swarm Creativity

It is an odd word, not one that you'll find in the parlance of most organizations, business or otherwise. But the idea of "swarm," especially in the context of creativity, promises to become more and more familiar to those seeking to stay on the cutting edge of innovation in the new century.

In July 2003, Boston experienced the power of swarming. Flash mobs had already burst into applause on the mezzanine of New York's Grand Hyatt, and they had whirled like dervishes in San Francisco. On Thursday, July 31, 2003, they struck in Boston. As Harvard bookstore employees watched, a few hundred people crowded into the greeting card section, holding instructions on what to say and how to act. They were looking for a card for their friend Bill. He lives in New York. Ten minutes later, the mob broke into applause, and then dispersed. "Bill from New York" is credited with creating flash mobs. The gatherings are coordinated through Web sites and chain e-mails, with the point of being pointless. While the first flash mobs strove to stay silly, they have since grown from an Internet curiosity to an increasingly widespread part of modern Web society.

Even politicians like 2004's U.S. Democratic presidential candidate Howard Dean have been drawn into the flash mob phenomenon to organize their political campaigns. Dean relied heavily on the World Wide Web to mobilize his grassroots supporters, to collect campaign money, and to fine-tune his political message by collecting direct feedback from supporters on his

Weblog. John Kerry, the 2004 Democratic presidential nominee, continued when Dean dropped out, collecting a large part of his more than $300 million campaign donations from grassroots supporters using his Web site. The Kerry campaign learned the lesson from Dean. Thus, irrespective of what you might have thought of Dean's politics, no one can deny the power unleashed by his campaign's "swarm."

Why should this matter to business executives?

Swarm Intelligence for Social Insects

Flash mobs are striking examples of how the Internet can be tapped to coordinate "swarming" behavior, a concept popularized by computer scientist Eric Bonabeau. Having studied the amazing world of swarm intelligence, the collective intelligence of social insect colonies, Bonabeau is now applying his insights to human interactions and computer technology.

Individually, one insect may not be capable of much; collectively, social insects are capable of achieving great things. They build and defend nests, forage for food, take care of the brood, divide labor, form bridges, and much more. Look at a single ant, and you might not think it is behaving in synchrony with the rest of the colony. But we sometimes observe "ant highways"— impressive columns of ants that can run from tens to hundreds of meters. They are highly coordinated forms of collective behavior.

Swarm intelligence in social insects is based on self-organization; no one is in charge, but social insects successfully solve complex tasks. According to Bonabeau, self-organization has four properties:[1]

1. *Positive feedback reinforces desired behavior,* such as when a bee recruits other bees to help exploit a food source
2. *Negative feedback counterbalances positive feedback,* such as when bees overcrowd a food source, which stops them from exploring it
3. *Amplification of randomness leads to positive reinforcement,* such as when bees that get lost trying to locate a known food source discover new food sources
4. *Amplification of interactivity has a positive outcome,* that is, when insects make positive use of the results of their own activities as well as those from the activities of other insects.

Social insects combine these four properties into predefined patterns, which have evolved over time, to efficiently accomplish a given task. For example, *colonies* of ants can collectively find the nearest and richest food source even if no *individual* ant knows its location.

Errors and randomness contribute very strongly to the success of social insects by enabling them to discover, explore, and exploit. Errors feed self-organization, creating flexibility so the colony can adapt to a changing environment with robustness, ensuring that—even when one or more individuals fail—the group can still perform its tasks.

Bonabeau reasons that our world is becoming so complex that it cannot be comprehended by any single human being. Swarm intelligence offers an alternative way of designing "intelligent" systems in which autonomy, emergence, and the ability to distribute tasks replace control, preprogramming, and centralization. During most of our history, we human beings have suffered from a "centralized mindset": we like to assign the coordination of activities to a central command. The self-organization of social insects, through direct and indirect interactions, is a very different way of performing complex tasks—and it closely resembles the behavior of collaborative innovation network (COIN) team members. Open source developers, for instance, exhibit behavioral patterns similar to those in an ant colony. While the behavior of the individual programmer might appear random, open source developers are, like ants, self-organized to build impressive software systems, directed by lead developers such as Linus Torvalds (the open source "ant queen"), who impress their distinctive brand and flavor on their "colony" of software developers.

The crucial point here is that in social insect colonies, as in COINs, there is no individual giving marching orders. Neither the ant queen nor the lead developer of an open source team directs the individual behavior of the individual.

The obvious advantages in accomplishing complex tasks through swarm intelligence include no central control, errors being okay, flexibility, robustness, and self-repair. If a decentralized, self-organizing system takes over, though, how should the individual behave so that the network performs appropriately at the system-wide level? It is difficult for conventional managers to accept the idea that solutions are emergent rather than predefined and preprogrammed. Most managers would rather live with a problem they cannot solve than with a solution they do not fully understand or control. The attitude of many managers to COINs reflects this view; managers resent giving up control in exchange for self-organization, increased agility, and increased flexibility of their organizations.

Watched individually, members of a COIN may appear, like individual ants, to behave erratically. Like an ant colony, though, the entire COIN operates as a highly efficient self-organizing community. Its members communicate in patterns structured as small-world and scale-free networks (which we will learn about in more detail in chapter 4). They share a common code of ethics that forms a sort of collaborative bond based on similar expertise and shared goals. In the appendixes, we will see how to make use of the e-mail "pheromone" trail of COIN members (much like the chemical trail ants use to communicate and collaborate) to spot COINs and to optimize their performance.

Lessons learned from the amazing world of swarm intelligence of social insects also apply to how members of COINs innovate. The ant or bee queen is not served because she orders her insects to do so, but because evolution has taught the insects that protection of the queen means protection of their gene pool for survival. The same is true for the people working together in a COIN. People working with the innovator are not working for her or him because they have been ordered to do so, but because they want the innovation to succeed. They all share the same vision and goals (in a sense, the same "genes"); they want to succeed, and they want to see their innovation spread and be accepted by the outside world.

What Is Swarm Creativity?

Perhaps nothing expresses the main motivation for swarm creativity better than something I read on a napkin in a San Francisco restaurant: "If you and I swap a dollar, you and I still each have a dollar. If you and I swap an idea, you and I have two ideas each." By openly sharing ideas and work, a team's creative output is exponentially more than the sum of the creative outputs of all the individual team members. While swarm intelligence is based on equal sharing of information, swarm creativity is founded on sharing ideas openly. COIN members share their common vision. They share all their findings and results of their common work, and they also share the credit for the results of their work. COINs collaborating in swarm creativity can achieve awesome results.

One of the most amazing examples of swarm creativity I know of is a virtual collaboration among three jazz composers and musicians, who play together at the same location but compose and record the same piece in parallel by coordinating over the Internet. Without ever getting together physically, they jointly composed and recorded an opera called *Popes*.

The three composers who succeeded in this team effort—David Hoenigsberg, Douglas Grannell, and Gareth Dylan Smith—met several years ago and spent considerable time playing and improvising together. They discovered their mutual compatibility. They shared common behavioral and musical patterns, a joint way to tackle a musical task, or, as Hoenigsberg calls it, a "common sound world." This is familiar for accomplished jazz musicians, who get together to play for hours on end without written melodies. They operate as a perfect team, where a theme is taken up and passed from one instrument to the other, with team members switching from playing in the background to getting the solo part.

Hoenigsberg, Grannell, and Smith extended the same concept into the cyberworld, virtually improvising together over long distances when they composed their joint opera. The three men were immersed in the same virtual world, collaborating as mutually aligned musicians, using the same musical and behavioral patterns. They produced an end product in which the sum is larger than its parts—a product none of them could have produced individually. In creating *Popes,* Hoenigsberg, Grannell, and Smith provided a perfect blueprint for a self-organizing, creative team innovating together and collaborating by sharing a common behavioral code.

It's notable that the three composers were not paid to do their work. They could not expect any immediate financial reward or immediate fame; moreover, their style of music is appreciated only by a relatively small group of dedicated lovers of modernist classical harmonic music. Nevertheless, they all shared the same musical goal, and they formed a highly motivated team that overcame considerable obstacles in creating an opera.

Swarm creativity needs enough leeway to flower. Team members want a risk-free work environment where they get mutual emotional support. Hoenigsberg, Grannell, and Smith work together as a team of equals, characterized by high mutual respect for each other's capabilities. When I interviewed them (individually), each spoke of the others with the greatest respect and tolerance for their strengths and weaknesses. Their relationship is based on a high level of trust, which has been built up by playing together over the years. The success of the long-distance *Popes* collaboration strengthened that trust.

Team members withhold no information. The free flow of ideas and thoughts is essential to the success of creative teams. While Hoenigsberg, Grannell, and Smith have a clear split of musical responsibilities, with a clear picture of which components each will compose, they stay in close contact. They exchange e-mails and phone calls frequently, bouncing inspirations and melodies back and forth and gauging each others' reactions to new ideas. This is not a process where there is always unanimous agreement. Rather, disagreements are raised openly and discussed. From this conflict, new creative inspirations arise.

This composer team has no external coordinators. It is entirely self-organizing and self-selecting. Roles and responsibilities of each member are clear to all, with no need for lengthy coordination meetings. There is no officially appointed leader, and Hoenigsberg has assumed a coordinating role only because, in his words, he "sort of naturally" inherited the function. He sees himself not as an authoritative boss, but as a servant of his team who does his best to help each team member succeed in his chosen task.

The *Popes* project, then, illustrates excellently the various facets of swarm creativity. However, it is only a three-person innovation team. There are COINs with tens of thousands of members engaged in swarm creativity. An ideal example is Wikipedia: it is the ongoing result of the creative output of more than 100,000 volunteer authors and editors.

Wikipedia—Swarm Creativity Thriving Online

Wikipedia, the online encyclopedia, thrives on swarm creativity. It is the collaborative, nonprofit competitor to commercial online encyclopedias such as Microsoft Encarta and *Encyclopaedia Britannica.*[2]

Wikipedia was founded in 2001 by Larry Sanger, a philosophy lecturer at Ohio State University, and Jimmy Wales, an Internet entrepreneur. While Encarta and *Encyclopaedia Britannica* are maintained by a staff of paid journalists, researchers, and scientists, Wikipedia's more than 400,000 entries are updated by thousands of volunteers. Wikipedia makes no distinction between authors and readers; any reader can change an encyclopedia entry at any time. The current directory of Wikipedians, as the volunteer contributors call themselves, lists more than 100,000 names, extended by an unknown number of anonymous authors.

Wikipedia entries are surprisingly accurate and complete because they can be corrected and updated. Authors are asked to maintain a neutral viewpoint (as well as "to be bold but stay cool" and to "be nice to newcomers") in order to keep the contents objective. A small group of trusted regular authors and editors are appointed as administrators because not everyone follows the guidelines and etiquette, or "Wikiquette." The strongest measure of censorship is that this small group can temporarily turn off editing of an entry for the rest of the world.[3]

Thus far, Wikipedia has worked remarkably well, based on a policy of nearly complete freedom to edit. The Wikiquette asks each Wikipedian to treat her or his fellow community members with respect. Wikipedia is entirely transparent in how text for controversial entries is fought over by different factions in the maintainer team. A log for each entry tracks the change history. Each entry also has a separate discussion page, where disagreeing authors are asked to engage in polite dialogue and where people should be prepared to apologize. An important part of the Wikiquette describes how to give praise when praise is due by making friendly entries on a user's page, by listing work under the category "great editing in progress," and by nominating entries as the day's "featured article."

Wikipedia is not the only COIN sharing the fruits of its collaborative endeavors at no cost. The open source movement is one of the best examples of a network of COINs, which innovate in swarm intelligence by contributing their intellectual property to the world and giving away their software source code for free.

Open Source Software—The Advantage of Swarm Creativity

One of the most compelling arguments for operating in swarm creativity is the astonishing rise of the Linux operating system. Finnish computer science student Linus Torvalds was certainly unaware when he began his term project to develop a UNIX-based operating system for the personal computer (PC) that Linux would become—in just a little over 10 years—the main competitor to Microsoft's dominance of the small server and desktop operating system (OS) market. The search engine giant Google uses Linux to power its farm of thousands of Web servers.

Linux has become the most popular alternative OS to Windows—not just because it is available for free, but also because anybody can make changes to it. The entire source code of Linux is publicly available, and anyone adept at the C programming language can extend the code. Torvalds asks only that improvements to Linux be shared; he and a group of volunteers check them and integrate the best into future versions of Linux.

What Torvalds created is *open source software,* which comes in many flavors but shares certain characteristics. One is that the source code—the underlying instructions that determine how a program operates—is public, so users can tamper with it; another is that new code is usually contributed by volunteers. So far, most open source projects are concentrated in areas of infrastructure software—the code that runs the core activities of a computer or a network. This extends from the basic operating system to the middleware layers of software needed to run particular applications.

The success of open source software projects illustrates the power of COINs and relates directly to some of the benefits introduced in chapter 1. Key to the success of Linux, for instance, has been *total transparency* and a set of consistent rules among thousands of programmers involved. Advocates of open source also claim—with considerable justification—that their software is of a higher quality than other software. Because there are so many users and developers of open source software products, there wil be no "deep"—that is, hard to find—bugs, or software glitches. This has been verified by several research groups.[4]

In the case of closed source software, users usually update only after software programmers have fixed reported bugs. A new version is available only after discrete periods of time, during which private programmers work to fix the last batch of faults. For open source software, this type of feedback and updating happens *continuously.* Open source software has a faster development and fix cycle compared to proprietary systems, and it also benefits from more user feedback. The same factors also help open source software develop stronger *security* protection, making open source software more secure than proprietary software.

The conclusion: open source software is developed faster, better, more securely, and at lower cost than closed source software. Teams of open source software developers operate as near-ideal COINs, coordinating their creative efforts through swarm intelligence. Their products are at least as good as those produced by conventional closed source programmer teams in conventional companies. At the same time, open source programmers also operate as a meritocracy, where the intellectual and emotional capabilities of individuals define their position in the open source development community.

IBM is one company that has discovered the advantages of COINs in combination with open source. The computer giant has invested more than $60 million into the open source tool set Eclipse, which has become the leading Java software development environment, attracting hundreds of vendors and millions of users. In fact, growth and adoption of Eclipse have been so rapid that companies such as Ericsson are evaluating it as a single, corporate-wide development platform to cut costs and boost efficiency.

In the beginning of 2004, IBM completely released its control over Eclipse and turned it over to an independent foundation. IBM gets to appoint just one of 10 members of the foundation's board of directors, but it continues to invest substantially in Eclipse by employing more than 50 Eclipse core developers with "committer" status—that is, people who form the innermost core of the Eclipse open source developer community, owning the rights to make changes directly to the source code.

Why does IBM invest in Eclipse under an open source license? I met with Erich Gamma, one of the lead developers of Eclipse, in his IBM office in Switzerland. First, he told me that IBM recognized that it cannot build all the software tools itself, that it needed a platform to help integrate different tools so that partners can contribute and extend the tool set. Obviously, such a platform also helps IBM align its internal tooling efforts; having a widely used software platform distributed under an open source license guarantees high quality, because it is tested and maintained by millions of programmers who discover, report, and fix bugs continuously. Second, IBM sells complementary software tools based on Eclipse, and having Eclipse available for free on almost any platform provides IBM with a huge potential customer base for these tools. Third, IBM reaps lots of fame, goodwill, and what Gamma calls the "cool factor" by being seen as a significant contributor of the best Java development environment to the public domain. And IBM is using Eclipse to develop some of its own commercial software packages, such as Lotus Workplace.

The Eclipse project provides a prime example of a COIN that has had a stellar success based on principles of meritocracy, transparency, and openness

—garnering numerous awards as the best Java product and Java development tool (see sidebar 2.1). Even more important, users have cast their vote, downloading the Eclipse tool more than 18 million times since its inception.

Intrinsically motivated and self-organizing into different roles of work, open source developers perfectly demonstrate the advantages of swarm intelligence (see sidebar 2.2). They produce results of superior quality at much lower cost than do commercial software developers. Working to develop new

Sidebar 2.1
The Eclipse Development Team Operates as a True Knowledge Network

The core innovation network for the Eclipse technology project is surrounded by learning and interest networks. Specifically, former Object Technology International development labs in Ottawa, Toronto, Portland, Minneapolis, Saint Nazaire, and Zurich (since acquired by IBM) comprise the network's backbone. Each of these labs hosts a small team of developers led by Eclipse project leaders; the leaders act as hubs of trust linking the core teams into a globally active COIN. These leaders have collaborated closely for many years and have established high mutual trust. The trust is then transferred to programmers working on teams at different locations.

Erich Gamma—an Eclipse lead developer who works at IBM in Switzerland—is one such hub of trust. He emphasizes *meritocracy, transparency,* and *openness to contributions from everyone* as the three basic principles governing the Eclipse team's work ethic. IBM employees receive no preferential treatment. New programmers, whom Gamma hires for his team at IBM Research in Zurich, must be confirmed and voted on by their Eclipse project peers, independent of company affiliation.

A new hire begins as a "debugger," writing software patches to fix bugs. Debuggers work their way up to become "developers," who actively participate in newsgroups and mailing lists. Next, they can be voted in as "committers," assuming they have proven their worth and gained a reputation as trustworthy programmers; committers gain write access to the source code repository and voting rights for new committers.

The project is managed by a project management committee (PMC), which today has three active members (the size has varied). New PMC members are admitted only after unanimous approval of all active PMC members (including Erich). In this way, decision processes regarding project management and tool development strategy are made transparent to the entire community. Membership in the Eclipse foundation is open to everyone.

Sidebar 2.2
What Motivates an Open Source Developer to Be a COIN Member?

Over a beer, Stefano Mazzocchi—the inventor of the open source Apache Cocoon framework—explained to me what inspires him. "There's a three-level pyramid that motivates open source developers. At the lowest level, open source programmers want to boost their ego by gaining a reputation as supreme programmers. At the next level up, there's the desire to gain new knowledge. At the highest level of motivation is the fun factor, of enjoying what they are doing."

Additionally, an online survey of members of the Linux developer community details three categories of motivation.[1]

1. *Collective* motives include identification of the developers with the goals of their community and the strong desire to help make those goals come true.
2. *Peer recognition* motives, not as important, include the wish to work together with friends and colleagues—and for those people to rank you highly.
3. *Direct reward* motives include learning new skills, making more money, or gaining new friends. Open source developers are motivated primarily not by money; having fun and learning ranked highest in the survey.

1. See www.psychologie.uni-kiel.de/linux-study/.

innovative products by massive collaborative creativity, they are powered by the self-organizing tenets of swarm intelligence: reinforcement of positive and negative feedback, and amplification of randomness and interactivity. It's a power that businesses can harness by supporting COINs.

More Examples of Swarm Creativity

I experienced swarm creativity firsthand when we created the Interoperability Service Interface (ISI) for Union Bank of Switzerland. We began by using public domain software. The project was running on a shoestring budget, driven by the dedication and commitment of the team members; these members had complementary talents, combining programming, system architecture, database, algorithms, management, and sales skills. The primary motivation for teamwork was clearly not financial, but to get the ISI system up and running, and membership was highly self-selecting. After all, while this was an opportunity to learn about the latest software technologies, it was a high-risk project and a huge technical challenge. Within the team, we worked in a high-trust environment with internal transparency. It was okay to have crazy ideas, which the entire team would discuss. In a few weeks, we were able to build a successful prototype system at very low cost, and ISI is still up and running. (The bank's COIN is described in greater detail in chapter 5.)

DaimlerChrysler's e-extended Enterprise (e^3) project, introduced briefly in chapter 1 (and detailed in chapter 5), also illustrates the properties and advantages of swarm intelligence in a COIN with obvious commercial goals. The project was a product of the e-business economy of the late 1990s, and so merely being part of this new and revolutionary undertaking was highly motivating for project team members. It was unbelievably exciting to participate in this journey into uncharted territory, at the leading edge of the field, at barely controllable speed. Nobody knew what the next day would bring. Unexpected developments and alliance opportunities popped up constantly. In those spring months of 2000, the entire team worked at 200 percent of its capacity, energized by the vision of its executive vice president, who let the team members take the reins and set the day-to-day directions themselves.

Clearly, the motivation here was not opportunities for raises and promotions, but the enjoyment of participating in a total overhaul and a melding together of the procurement processes of Daimler and Chrysler. This excitement was enough to overcome widespread initial skepticism of the management of the procurement and supply departments at both firms involved in the merger. Department heads on both sides of the Atlantic initially were reluctant even to share simple information such as the head counts of their business units, but their progressive buy-in into the shared vision of creating something totally new resulted in one unified COIN team. The outside torrent of e-business also created an environment of creative chaos and uncertainty that greatly increased the readiness of executives to accept change.

The project team worked together as a real collaboration network. Once the initial reluctance to share knowledge was overcome and mutual trust had been developed, information was made transparent to all team members. Although the team was physically split between locations in Stuttgart and Detroit, team identity was quite strong. Having monthly face-to-face workshops alternately in Germany and the United States helped build a shared identity and a common code of ethics; in those close interactions, strengths and weaknesses of each team member became obvious to the entire team, leading to an open and honest communication style. Team members, regardless of their rank in the company, were not afraid to speak out to the entire group. Complex problems were broken down into tasks and assigned to team members on both sides of the Atlantic. Thanks to a shared common understanding, multifaceted issues such as the development of a new supply chain model or the evaluation of supply chain automation software tools could be solved efficiently by team members working in parallel in both locations.

Membership in the team was self-selecting. Some team members who were initially appointed by senior management pulled out voluntarily, while others who initially were just seconded for a certain presentation or a specific task became increasingly engaged and asked to be officially transferred into the project, subsequently becoming mainstays of the team.

In the end, the e^3 project was a big success by most measures of project management. Not only were most deliverables on time, but they were also developed efficiently and at reasonable costs. New procurement processes and software solutions were successfully deployed, and the entire project more than paid for itself in the first year of operations. Project team members moved on to assume prominent roles within DaimlerChrysler's procurement organization and were even chosen for leadership positions with outside firms.

Again, the business success of this project is undisputable: the company spent slightly more than $10 million to optimize its procurement function and saved more than $50 million in the first year of operation. And it succeeded because of *collaborative innovation.*

Why Collaborative Innovation Is So Powerful

What starts a COIN, and what triggers successful innovation? Is it the phase of the moon, or just sheer luck, that is responsible for creative inspirations and the emergence of groundbreaking new ideas? The crucial point about innovations developed in COINs is that they are open and disruptive.

Harvard professor Clayton Christensen distinguishes between sustaining and disruptive innovation.[5] *Sustaining* innovations do existing things in a novel and better way. For example, Wal-Mart revolutionized mass retailing by using information technologies in a much more massive and better-integrated way than its competitors. It optimized its supply chain by pioneering concepts like vendor-managed inventory and automated demand forecasting and re-plenishment, beating its competitors and becoming the number one Fortune 500 company globally, as well as one of the most successful enterprises by many measures. But the "innovation" of Wal-Mart consisted of doing the same things that its competitors did, only "better," much more efficiently. For example, Wal-Mart used available information technologies more thoroughly and consistently than its competitors by being one of the first companies to let its suppliers manage shelf space in Wal-Mart's own stores.

On the other hand, *disruptive* innovations such as the World Wide Web and Linux, the airplane, and the railroad totally changed the way in which something was done.[6] Disruptive innovations are normally conceived outside of established organizations or, if they come out of organizations, they are developed without organizational blessing and are not aligned with the official organizational vision and goals. They have the potential to turn established organizational norms upside down and to redefine the way people work and live.

Academics such as Christensen and Eric von Hippel have analyzed how disruptive innovations succeed.[7] Their main insight is that disruptive innovations rarely become embedded into large enterprises. While they might initiate within large organizations, they succeed outside the formal hierarchies of huge enterprises. In fact, the teams that take new inventions outside of the large corporations to start new companies frequently initiate as "stealth" COINs inside large corporations.

In the past, the centrally funded research labs of large corporations were responsible for many disruptive innovations. AT&T Bell Labs (now part of Lucent Technologies), IBM Research's labs, and the Xerox Palo Alto Research Center (PARC) produced the transistor, speech recognition, copper chip technology, the scanning tunneling microscope, and the Ethernet; they all fundamentally changed the way a certain product or process operated. But the closed innovation model, in which companies fund these central labs to develop new technology and products, is no longer generally valid. What Harvard professor Henry Chesbrough calls the *open* innovation model is the new way in which individuals, small companies, and innovation networks create innovations.[8]

The obsolescence of the closed innovation paradigm has been accelerated by different factors. The explosive growth of a large base of skilled, mobile workers has created an auction market that allows startups and entrepreneurs to hire "the best and the brightest" away from larger companies, which threatens the bigger firms' ability to sustain research and development (R&D) investment. Between 1980 and 2000, U.S. venture capital skyrocketed from roughly $700 million to over $80 billion. Ready access to this capital, in combination with lucrative stock option packages, attracted many key lab personnel from large companies to startups, eroding the knowledge bases of some larger companies. The availability of venture capital, worker mobility, and global communication means that research employees no longer have to keep their ideas on the shelf while they wait for internal approval to explore them. Instead, employers' rules permitting, they can commercialize those ideas independently. External suppliers are becoming more proficient at delivering products that are just as good, if not better, than what an internal corporate R&D lab can develop—a trend that levels the playing ground between large firms and their smaller competitors. It allows large firms to bring new products to market faster, while giving smaller rivals the opportunity to overtake larger companies. To take advantage of the changes brought by those factors, companies must move to the open innovation model that leverages the knowledge base of COINs. If they don't, COIN members will leave and start their businesses elsewhere!

Disruptive "grassroots" innovations by COINs, such as the Internet, the Web, and Linux, grew out of research labs and universities with little or no organizational blessing and with minimal budgets. The main drivers of these innovations were not the compensation-and-reward systems of large corporations, but the dedication and commitment of the researchers and innovators. IBM, Intel, Microsoft, and other companies have learned this lesson and are successfully applying the same principles of open innovation to their own internal research processes. They have intensified collaboration with university researchers and small knowledge companies, and they take new innovations to market much earlier than in the past. A more collaborative culture combined with open communication has led to increased innovation, ultimately resulting in radically new revenue opportunities for companies that embrace the open innovation paradigm.

Open disruptive innovation is not restricted to the high-technology area. For instance, the people spearheading snowboarding are typical open disruptive innovators (see sidebar 2.3). They are pioneers—initially frowned upon by the majority. They are not afraid of riding off-slope, and they are developing their own solutions to manage their journey into new territory until their inventions become an accepted part of daily life. The commonality

Sidebar 2.3
Snowboarding as a Disruptive Innovation

The meteoric rise of snowboarding in just over 30 years is an excellent showcase for open disruptive innovation in a non–high technology setting.

The sport's earliest innovator was Jack Burchett, commonly credited with building the first snowboard in 1929 when he cut out a plank of plywood and tried to tie his feet to the plank with some clothesline and horse reins. Sherman Poppen, a chemical gas engineer in Michigan, took up the idea 35 years later, inventing the "snurfer" for his daughters by securing two skis together and putting a rope at the nose that the rider could hold onto to keep the board more stable. The snurfer was reasonably successful, and Poppen started a company to market his invention. But the snurfer was only seen as a toy for kids. It wasn't until the 1970s that snowboarding really took off. Dimitrije Milovich, a college student in upstate New York, worked as a waiter and was sliding down hills on cafeteria trays. He and other enthusiasts began to build snowboards that combined the concepts of surfboards and skis.

In 1982 the first U.S. National Snowboard race was held in Suicide Six, outside Woodstock, Vermont. Three years later, although a mere 39 of the approximately 600 U.S. ski resorts allowed snowboards, the craze had spread. St. Moritz, Switzerland, hosted the first snowboarding world championship in 1987. By 1994, snowboarding was an Olympic sport.

The early snowboarding pioneers wanted nothing to do with the ski establishment, and the feeling was mutual: most skiers frowned upon the brave souls who rode their own devices and the first snowboards. Today, though, snowboarding

---

between the inventors of snowboarding and high-technology pioneers like Tim Berners-Lee is their combination of capabilities as team players with creative intelligence.

Fritz Bircher, a professor of electrical engineering and computer science at the University of Applied Sciences of Burgdorf, in the Swiss canton of Bern, is an example of the kind of *creator* who combines emotional intelligence with creativity and proves an ideal influencer for a COIN. In fact, he is the archetype of a creator—constantly on the lookout for new, interesting applications for his latest invention, the MobileJet, an inexpensive inkjet printer on wheels that can print arbitrarily large areas on almost any surface. Bircher started the MobileJet project in 2000, when professors at another Swiss vocational college asked one of their Burgdorf colleagues for help. Now, as we met, he

is a decidedly mainstream activity, and snowboarders are as welcome as skiers on the slopes.

Sherman Poppen's "Snurfer." 1960s Brunswick Snurfer ad courtesy of Brunswick.

was getting a call from an investor who was interested in commercializing his invention. But Bircher's motivation is to design and build new things; making money on his invention is of secondary importance.

Bircher is currently working with another Burgdorf professor, Reinhold Krause, who is an expert in mechanical engineering. They have assembled a team of about a dozen research assistants to bring their ideas to fruition. Bircher talks constantly about how his team's members come up with solutions by working together; in this aspect, he is a typical leader of a COIN, where people work together in a meritocratic environment driven by a common vision and common goals—not for short-term profit. Krause and Bircher form the perfect core team of a COIN. Both are driven by high intellectual curiosity, and their motivation to innovate is definitely not financial. By fostering an open collaborative and meritocratic culture in which credit is given where credit is due, results are produced not because people are paid to deliver them, but because people care about what they are doing. This leads to a continuous stream of new innovations. For instance, the MobileJet printer is being extended to work as an inexpensive and efficient Braille printer. Krause and Bircher are also talking to artists about using the MobileJet to produce large-scale works of art on concrete and tarmac surfaces. These two men have firmly embraced the open innovation model, and they collaborate and share their results with other innovators at universities and in industry research labs.

Krause and Bircher made headlines with another invention a few years ago. Their global positioning system (GPS)-controlled lawnmower, with autopilot, cut grass to an accuracy of two centimeters. This invention led to a whole list of innovations: building the GPS system into huge combine harvesters; using the same technology in a maintenance robot to measure and maintain Swiss railway tracks; and adapting the technology to road rollers, increasing the efficiency and more evenly compressing asphalt coverings in new road construction.

Krause and Bircher are creators and "gurus" exhibiting the right mix of creativity, charisma, and collaborative skills. They excite other people about their innovations. There are other types of influencers who are crucial bringing a new innovation over the tipping point to success. Let's look at one of the most famous innovators of all time—Leonardo da Vinci.

Innovation by Collaboration

In addition to being one of the most creative innovators ever, Leonardo da Vinci was an excellent team player. His talent for dealing with other people in difficult situations was legendary. His biographer, Giorgio Vasari, describes how, when da Vinci was still an apprentice with the famous painter Andrea del Verrocchio, his depiction of an angelic figure was so masterful that Verrocchio considered giving up painting. Da Vinci replied that it was the greatest compliment to the master that the student should exceed the master's ability.[9]

Later in life, da Vinci built up a community around him that included renowned mathematician Luca Pacioli, as well as devoted younger artists such as Andrea Salai and Francesco Melzi. Vasari wrote that da Vinci's "generosity was so great that he sheltered and fed all his friends, rich and poor alike, provided they possessed talent and ability."

People like Leonardo da Vinci are lifelong learners, "swarming together" with other intellectual giants. Pacioli, for example, not only invented double accounting, but also worked with da Vinci on his engineering exploits. Da Vinci purchased a copy of Pacioli's famous book on Euclidian geometry before the two ever met, and he drew the figures in their joint work *On Divina proportione,* a book in which they studied the "golden ratio" and related Euclidean theorems. Later, da Vinci and Pacioli taught each other in a mutually enriching learning and tutoring relationship, with da Vinci clearly having the sponsoring role.

Leonardo da Vinci is arguably the most famous inventor ever. But it makes no sense to look at innovators in isolation. We need to study both the creative genius and the people that surround him. On the one hand, the innovator needs to be practical and down-to-earth enough to be understood by his surroundings; innovators must be able to position their inventions in their environment and successfully convince others of their value to create sustainable progress. That's what da Vinci was able to do with his mechanical engineering and military inventions and his paintings, frescoes, and statues. On the other hand, the innovator must also get along with the people surrounding him. Each innovation needs a group of early adaptors and rapid followers that understand the genius innovator and his innovative ideas. They form the beehive, or ant colony, that acts as bridge and translator to spread the concept. At the core of the COIN are the charismatic, inspirational, and creative thought leaders who are enough in synch with their time and environment for their innovations to be recognized as such. But similarly important are the people around the core innovators, people who are willing to play a supporting role in making a great idea succeed.

While we all have the preconceived image of the genius in his experimental cabinet cranking out inventions, the people surrounding the genius are at least similarly important, forming the COIN working in "swarm creativity" to develop the innovation. Leonardo da Vinci had a whole support organization of painters, sculptors, and other artists on his payroll that helped to deliver his master works. Da Vinci treated and paid them well; in return, they worked for his "brand." Da Vinci would not have succeeded had he not spoken in a language understood by his environment and had he not been surrounded by a team of collaborative innovators. Of course, even da Vinci

---

Sidebar 2.4
Swarm Creativity at Work in My First COIN

My experience with COINs began when I was a postdoctoral researcher at the MIT Laboratory for Computer Science. I was part of a small group of people that wanted to do something revolutionary for computer science education. Our goal was to develop an educational multimedia CD-ROM that would convey in a more interactive and visual way the highly complex subject of computer algorithms—one of the hardest parts of computer science. We were not the first researchers to produce computer animations for algorithms, but we were the first seeking to build an integrated system.

We enjoyed two types of support. First, three MIT educators who had written a popular book on algorithms gave us their full text, allowing us to base our animations on their explanations and pictures.[1] Second, and more important, one of these coauthors agreed to sponsor a class I was planning to teach at MIT about algorithm animation and hypertext. We were able to build a prototype system of "animated algorithms" as a series of class projects, and that convinced our lab director to give us a small grant to turn the prototype into a full commercial product. This enabled us to launch our little COIN, comprising a highly motivated core team (working basically for free) and a team of a dozen students. We used the grant to buy some hardware and to compensate students for refining the algorithm animations produced in the course.

---

sometimes fell into the trap of being too much ahead and not communicating in a language understood by his disciples. When he described parachutes, planes, and helicopters, for instance, he was ahead of his time by hundreds of years (figure 2.1).

We could say that da Vinci was the core of a self-organizing swarm of similarly minded artists and scientists, working together in the creative chaos of Renaissance Italy. The swarm used the same self-organizing properties social insects employ: positive feedback, negative feedback, randomness, and multiple interactions. It was a nurturing environment where collaboration was amplified by positive feedback (for instance, da Vinci and Pacioli working on their book on golden ratio, which neither could have completed on his own). Early on, da Vinci honed his collaborative skills when he successfully dealt with negative feedback by other artists less brilliant than himself.

The core team consisted of three people with almost no money but a great deal of enthusiasm and excitement—that was enough to keep us motivated. The students worked very hard, turning class project animations into production-level computer visualizations stable enough for a commercial product.

After another year at MIT, I returned to Switzerland. Some of the students moved to other universities. But while we were now geographically distributed, the team remained on track. After two years of hard work our COIN published a CD-ROM of *Animated Algorithms* to excellent notices from reviewers and users alike.[2]

This COIN experience showed that COINs are both powerful examples of teamwork and great levers of individual strengths. The project would have been impossible for any one of us working alone, but together our swarm creativity was unbeatable. Our success was possible only because of the team spirit that had developed in the COIN, where the whole team was driven to realize a shared vision. The core team acted as creators, as collaborators who pulled together a diverse team and kept it hard at work, and as communicators who convinced sponsors and coworkers of the project's merits. And while we shared a common set of characteristics, our *individual* strengths were complementary skills. We worked hard not from a desire to make money, but from commitment to our shared goal and a desire to succeed.

1. Cormen, Leiserson, and Rivest, *Introduction to Algorithms* (1990).

2. Gloor, Dynes, and Lee, *Animated Algorithms* (1993).

Similar to the positive use of chaos and randomness by insects, da Vinci used the chaotic environment in northern Italy to maintain his independence, to find the most generous and venturous sponsors, and to gather around him the best disciples. Finally, similar to ants that need multiple interactions with each other to complete their tasks successfully, da Vinci could excel only in an environment bustling with other painters, sculptors, mathematicians, and scientists building upon each other's work in continuous "coopetition."

The diffusion of disruptive innovations generally follows the same process. First, there is a genius with a brilliant idea. Working in relative isolation, the genius develops the idea and implements the first prototypes. But such ideas are so far ahead of their time that the people around the genius do not recognize their greatness. It takes a collaborative leader able to congregate a

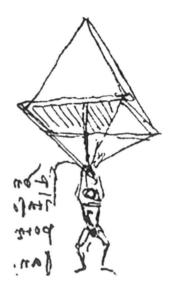

Figure 2.1.   Da Vinci's parachute design.
Copyright © Biblioteca Ambrosiana, Milan.
Codex Atlanticus, f.1058v.

group of dedicated disciples to spread the message. This leader has a charismatic yet collaborative personality, is highly persuasive and imparts contagious enthusiasm; he or she shepherds disciples toward the communal vision. The motivated disciples infect the environment with their own enthusiasm and keep things going until they reach the tipping point, where the new idea becomes part of daily life. Together, the collaborative leaders and the motivated disciples constitute a COIN.

Familiar Examples of the COIN-Driven Innovation Process

There are many familiar examples of this COIN-driven innovation process, even if readers may not realize the involvement of COINs. PCs, the Web, the Internet, and Linux have already been mentioned. There were early innovators such as Douglas Engelbart, who returned from World War II, where he had been a radar technician, with a vision of working interactively with computers. Engelbart's inventions were crucial for the success of both the personal computer and the Web. Between 1963 and 1968, he invented technologies to support individual knowledge workers and collaboration within groups of knowledge workers. Out of his work came the computer mouse, multiple windows on a screen, and e-mail. Another researcher, Ivan

Sidebar 2.5
Torvalds Takes on the Critics of Open Source

Linux pioneer Linus Torvalds famously cited another great innovator in arguing for the openness and collaboration intrinsic to COINs—the core method of open source software development. Responding to statements by Craig Mundie, a Microsoft senior vice president who criticized open source software for being non-innovative and destructive to intellectual property, Torvalds sent out this e-mail retort:

> I wonder if Mundie has ever heard of Sir Isaac Newton? He's not only famous for having set the foundations for classical mechanics, but he is also famous for how he acknowledged the achievement: "If I have been able to see further, it was only because I stood on the shoulders of giants." . . . I'd rather listen to Newton than to Mundie. He may have been dead for almost 300 years but despite that he stinks up the room less.[1]

1. See Wikipedia at http://en.wikipedia.org/wiki/Linus_Torvalds.

Sutherland, created the first graphics package, called "Sketchpad," while working at the Massachusetts Institute of Technology (MIT). Sutherland and Engelbart were typical "early genius" inventors, defining a new field with no expectation of raking in financial profits.

Collaborative leaders of the PC revolution include Alan Kay, Butler Lampson, Robert Taylor, and Charles Thacker. They built the Altos, among the first personal computers, while at Xerox PARC. The first of the motivated disciples to clearly recognize the significance of that invention was Steve Jobs, whose Apple Computer Company popularized the concept of the personal computer. Later disciples collected even larger rewards. People such as Robert Noyce, cofounder of Fairchild Semiconductor and Intel, and Bill Gates, cofounder of Microsoft, combine technical skills with uncanny foresight and vision, the ability to create a highly cohesive company culture, and great salesmanship. We can see how well this has repaid them!

The creation of the Internet, the global computer network that is the foundation of the Web, delivers an even more compelling story of how a group of intrinsically motivated people can drive open disruptive innovation.

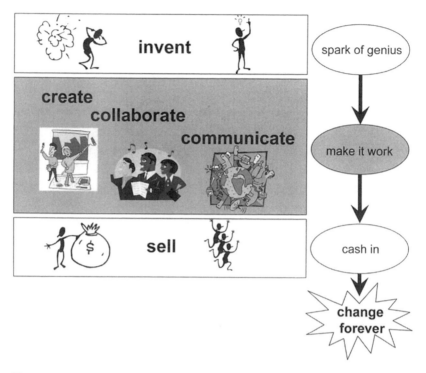

Figure 2.2.   COIN-driven innovation process.

The Internet's creation was very much driven by COINs. Soon after the first mainframe computers had been developed for military use in World War II, a few visionary thinkers realized that, by linking these mainframes, they could make much better use of computers. They were motivated not by financial reward or promotion, but by a deep desire to further technical development. It was clear to those visionaries that the usefulness of the computer network would grow exponentially with the number of linked computers.

The goal of the early Internet visionaries was to develop an open networking standard linking computers of any make and model anywhere on the globe using publicly available networks such as the phone system. J. C. R. Licklider, or "Lick," as friends, colleagues, and acquaintances called him, was the best known among the pioneers; he had a vision of allowing multiple users to simultaneously use the expensive mainframe computers of his time. Newly appointed in 1962 to head the Computer Research Division at the U.S. Defense Department's Advanced Research Projects Agency (ARPA),

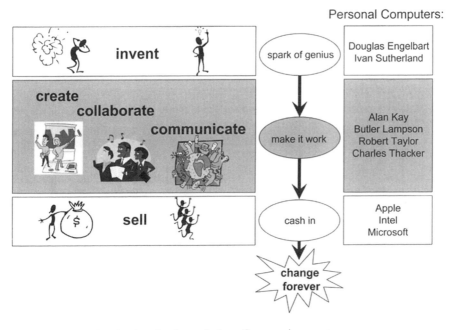

Figure 2.3.   COIN leaders for the evolution of personal computers.

Licklider oversaw the biggest computer-related research budget of the U.S. government. Against the advice of the computer establishment, he used his position to mastermind time-sharing computers and then to fund development on the Advanced Research Projects Agency Network (ARPANET), thus single-handedly creating internetworking. In 1960, he accurately foresaw the Internet, envisaging "thinking centers, that will incorporate the functions of present-day libraries together with anticipated advances in information storage and retrieval. . . . The picture readily enlarges itself into a network of such centers, connected to one another by wide-band communication lines and to individual users by leased-wire service."[10] Contemporaries describe Licklider as a man who retained his modesty despite his considerable influence on computing. The list of people he selected and put in charge of implementing his vision at research universities that included MIT, Carnegie Mellon, the University of California at Berkeley, and Stanford reads like a "Who's Who" of today's computer research leaders.

The people who further extended the ARPANET into the Internet formed a COIN in its purest form. A self-organizing network of individuals, the

Sidebar 2.6
The Dedicated Disciples Who Carry Innovation over the Tipping Point

In his best-selling book *The Tipping Point,* Malcolm Gladwell describes the characteristics of the dedicated disciples it takes to make a great idea a great success.[1] These *connectors, mavens,* and *salespeople* are a crucial part of the process.

The *connector* knows many different people in different socioeconomic areas. For example, Paul Sachs, an heir to the Goldman Sachs banking empire, was a connector. When the Museum of Modern Art (MOMA) was founded in New York in 1929, Sachs played a crucial role, spanning many of what social scientist Ronald Burt calls "structural holes" to connect the modern arts world and New York high society's wealthy benefactors. Later, Sachs became an art professor at Harvard University and recommended one of his best students, Alfred Barr, for the job of first MOMA director. He also assisted the museum's founders in finding potential sponsors, board members, and works of art. Paul Sachs was an archetypical connector, linking the distinct fields of high society, finance, and modern art.

Mavens are quite different from connectors. They do not necessarily have to know many people, but they are experts in a certain field and—driven by altruism—they love to provide this expertise to others. Mavens are helpful "connoisseurs" who carry the message of a great innovation because they are convinced it is great. The very nature of this supportive task makes mavens much less visible (usually) from the outside. We encounter small mavens when we get a tip about the best restaurant in a certain neighborhood, or when someone helps us connect complicated technical equipment.

We have all met the *salespeople* who comprise Gladwell's final category. They can sell just about anything—not through pressure, but because they have "emotional intelligence."[2] Real salespeople feel what their clients want and are able to package and present their products in a way that profoundly appeals to those clients. Great politicians are usually excellent salespeople; both Ronald Reagan and Bill Clinton had an uncanny sense for the concerns of the "common man." They stayed in close touch with the electorate while in the White House. This had a lot to do with why they won second terms and enjoyed unprecedented job approval ratings.

1. Gladwell, *The Tipping Point: How Little Things Can Make a Big Difference* (2002).

2. Goleman, *Emotional Intelligence: Why It Can Matter More than IQ* (1997).

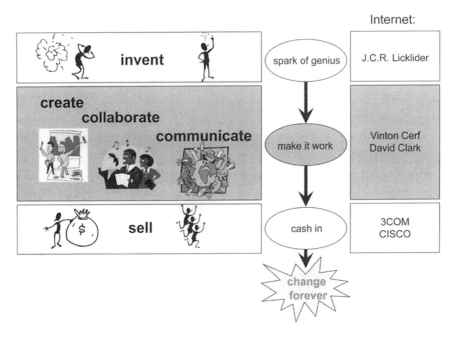

Figure 2.4.   COIN leaders for the creation of the Internet.

Internet Engineering Task Force (IETF) and its predecessor organizations have steered the development of the Internet for the past 30 years—without any central authority. Anybody could (and still can) join an IETF working group.

> The Internet Engineering Task Force is a loosely self-organized group of people who make technical and other contributions to the engineering and evolution of the Internet and its technologies. . . . There is no membership in the IETF. Anyone may register for and attend any meeting. The closest thing there is to being an IETF member is being on the IETF or working group mailing list.[11]

There have been hundreds of IETF working groups since its first meeting on January 1986 in San Diego, California, with 15 attendees. The predecessor of IETF, the Network Working Group (NWG), goes back 18 years earlier. Those working groups have been developing seminal standards and public domain implementations like transmission control protocol/Internet protocol

(TCP/IP), the main communication protocol of the Internet and Internet mail, and the protocol that manages millions of hosts on the network. Each IETF working group functions as a COIN, where a small core team—driven by visionary leaders such as Vinton Cerf and David Clark—collaborates on developing the specifications. Meanwhile, other members act as a sounding board, reading and commenting on the specs and testing the so-called "reference" software implementations developed by the core team. IETF working group participants receive no financial compensation. They are normally on loan from a company research lab, are paid by an academic institution, or are self-employed consultants. Their main motivation to participate is the desire to be part of the further development of the Internet and to advance the state of the art in internetworking technology.

These individuals share this intrinsic motivation with the group of programmers that created the World Wide Web, which evolved in a similar way to the Internet. Early geniuses such as Vannevar Bush, Ted Nelson, and Douglas Engelbart laid the groundwork by inventing technologies needed for the Web's success, such as linking, hypertext, multiple windows, and the computer mouse. Collaborative leaders such as Tim Berners-Lee and Robert Cailliau recognized the importance of these inventions and put the pieces together. But, again, it took motivated disciples to carry the innovation over the tipping point and trigger the tidal wave of e-business. Netscape's Marc Andreessen, e-Bay's Pierre Omidya, and Amazon's Jeff Bezos all combine the technical prowess of a maven with the wide people network of a connector and with great salesmanship. Not only did they launch highly successful companies, but they also redefined their respective industries and reshaped their competitive landscapes in the process. They share these characteristics with the open source programmers who created Linux.

Collaboration between Innovators Is the Key

Fritz Bircher and Reinhold Krause in Switzerland are perpetual innovators. That makes them perspicacious COIN members. But besides being very smart and having creative ideas, they are also excellent team players. This means that, in their COIN that formed around GPS-controlled lawn mowers and MobileJet printers, they have built up a collaborative team culture, based on

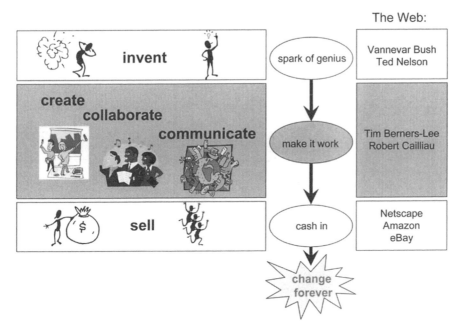

The Web:

Figure 2.5.   COIN leaders for the creation of the Web.

meritocratic principles. This does not mean that hierarchy is absent; to the contrary, it is clear that Bircher and Krause are the leaders. But they give credit where credit is due, and they foster a transparent work environment with high respect for the individual.

For the success of a COIN, it is the middle phase of the *invent-create-sell* process that is most critical. The MobileJet would never have been possible if Bircher and Krause had not collaborated creatively. In this second, collaborative creation phase, Krause's mechanical engineering skills and Bircher's software and electrical engineering skills were pooled together to carry a truly innovative product from concept to implementation, transforming it from a lofty idea into a working solution. The interaction between Krause, Bircher, and the 12 graduate students and research assistants in their COIN at Burgdorf was the catalyst in this crucial phase. Once they had something to demonstrate, Bircher and Krause communicated with the outside world. Their innovations have been featured on Swiss television, for instance, which has in turn led to contacts from potential users and investors.

47

For a COIN to succeed and get its ideas off the ground, there must be creativity, collaboration, and communication. Only a combination of people who bring these traits—the DNA of COINs—to the team can succeed in bringing a COIN's innovation over the tipping point. That DNA is the subject of the next chapter.

CHAPTER 3

The DNA of COINs

Creativity, Collaboration, and Communication

About 30,000 years ago, something amazing happened in Europe. For the previous 15,000 years, modern *Homo sapiens,* our ancestors, and Neanderthals had been living together. But then, in a prehistoric competition for survival, the Neanderthals disappeared. *Homo sapiens,* the survivors, went on to experience a cultural revolution, making great progress in art and craftsmanship.

Why did this happen? While Neanderthals were living in the same caves for generations, roving bands of our ancestors followed the herds, traveling great distances. During their travels, they met other people, exchanging ideas and learning about new innovations. Harsh external living conditions brought by the last Ice Age in Europe forced *Homo sapiens,* and those they met, to cooperate and innovate. While this external stimulus led to a flurry of innovations by *Homo sapiens,* it also led to the extinction of the Neanderthals.

So *Homo sapiens* beat out Neanderthals because of *learning networks*— this is the first of the five key elements of the genetic code of collaborative innovation networks (COINs).

Over the next 20,000 years, modern *Homo sapiens* progressed tremendously. The prehistoric trading camps grew into large cities housing thousands of people all around the globe, from Thebes in Egypt, to Ur in Mesopotamia, to Chang'an in China, to Rome in Italy. Beginning in Roman times, the Chinese and Roman empires were connected by the "Silk Road," trading goods and knowledge for over a thousand years. But with the demise of the Mongolian and Byzantine empires toward the end of the European Middle Ages, traffic on the Silk Road came to an end. One of the last travelers of the Silk Road was also one of the most famous: Marco Polo carried back to Europe knowledge about the Mongolian Imperial Chinese court, and about Chinese culture and daily life such as the Imperial postal system and paper money. And while the Silk Road was closed down because local warlords were fighting with each other, Renaissance Europe experienced a boom in science, culture, and commerce.

Almost a contemporary of Marco Polo, German Renaissance banker Jakob Fugger rose to become the richest man of his time. At the height of its wealth, the Fugger banking empire owned a tenth of all money flowing in the Holy Roman Empire, which stretched from today's Poland, Germany, Italy, Netherlands, Spain, and Portugal to Latin America and the Southeast Asian spice islands. While Emperor Charles V rightly claimed that in his empire the sun never set, he did not openly proclaim that it was Fugger who put him on the throne.

Jakob Fugger was not just an uncommonly shrewd businessman. He was also a great benefactor to the poor and needy and took excellent care of his own workers. He founded hospitals and elderly homes for his own employees, and he gave freely to the church and to the poor. In his hometown of Augsburg, he constructed the first social housing project, a subsidized apartment village for needy families who had to pay only a nominal rent. The Fuggerei still exists today, and it is still financed by the Fugger family; likewise, the Fugger private bank still operates, and the Fuggers are still a very wealthy family. Other European bankers and barons who competed with Fugger's bank for the favor and the purse of the emperor are gone; they cared only about their own well-being; and rarely did their fortunes significantly outlast the founding generation.

Sustainable business success requires a combination of keen business sense with *sound ethical principles.* That is the second key element of the genetic code of COINs. The third is *trust and self-organization,* the very things upon which Renaissance Europe thrived.

In 1449, Jakob Fugger and another rich merchant from Augsburg named Bartholomäus Welser financed the prospecting expeditions of Pedro Alvarez Cabral to Brazil, enlarging their immense fortunes in the process. At the same time, the Chinese emperor burned his huge fleet and decided not to venture beyond his mainland empire. And the diffusion of the invention of gunpowder provides a fascinating example of how xenophobic isolation can be disastrous. The Arabs and Chinese had invented gunpowder centuries before medieval scientist and Franciscan friar Roger Bacon first documented the formula in Europe, but the Chinese were not able to put gunpowder into productive use. They used it for fireworks, and even built crude gunpowder weapons, but the development of firearms in China stagnated until the Chinese encountered them again in a stunning defeat by the invading Europeans in the 19th century. Meanwhile, the warring city-states of tiny Renaissance Italy provided fertile grounds for nurturing the invention of the firearm. Charismatic Italian Renaissance potentates like Lorenzo di Medici or Pope Alexander VI Borgia, who relentlessly fought each other for political supremacy, provided huge incentives for creative innovators. The simple explanation for why China, a hundred times more powerful, stagnated in developing this sophisticated technology is that Chinese isolationist policy did not provide for the provoking, stimulating and at times chaotic hotbed environment that European Renaissance inventors such as Leonardo da Vinci faced. What the Ice Age did to Neanderthals, isolationalist policies did to imperial China. Satisfied with the status quo, the Chinese were overtaken by more nimble, innovative, creative, and chaotic competitors. China was ruled by an authoritarian emperor and a strictly regulated hierarchy of Confucian magistrates, while Italy and Germany were ruled by "creative chaos" and self-organization. People swarmed around petty rulers, bankers, and merchants acting as patrons of artists and innovators who autonomously took hold of their own destiny.

In Europe, a system of medieval guilds and cities took care of their own guild members and citizens; they created a system based on *self-organization* and *trust.* Nothing of the sort existed in China.[1] In Europe, craftsman guilds, merchant networks, courts of the gentry, and town communities functioned as interpersonal trust networks, providing a flexible foundation to develop new inventions at low costs for society. To survive, innovators had to collaborate and share what they knew within their trust networks.

The fourth key element in the genetic code of COINs, *making knowledge accessible to everyone,* is well illustrated by another European example. By the fifteenth century, the Catholic Church and the Pope—trusted spiritual leaders and the heart of the "barbarian" Western countries for over a millennium— had become a rampant hierarchical ulcer. Trust was perverted by an obscure and complicated hierarchy, individual greed, and corruption. Along came Martin Luther, who started his career as a simple priest in Germany, far from Rome. His Reformation began with questioning of the idea that timely salvation could be brokered by the church in the form of indulgences. Such egalitarian, anti-authoritarian habits of thought led him to more general assertions, finally giving ordinary people direct access to the word of God without the need of interpretation by priests. Luther sought to shift the power of spiritual authority back to where he felt it belonged: the hands of the flock. His ideas spread like wildfire. Today, Luther's "protestant church" has split up into smaller entities, becoming more transparent and operating by "swarm intelligence"—without a single, pope-like leader, where community members contribute their time and skills based on individual motivation in self-organizing knowledge networks.

The success of the Reformation was driven to a large extent by Johannes Gutenberg's invention of metal type. In 1455, Gutenberg published the Gutenberg Bible in Mainz, Germany. For the first time, the Bible became affordable not only to the rich, but also to the middle class. There was still an obstacle, though—it was printed in Latin, and the ordinary layman needed the clergy to translate the Word of God. Martin Luther changed this when he translated the Bible from Latin into German; his translation was first printed at Wittenberg by Hans Lufft in 1534. It is said that Lufft sold over 100,000 copies of Luther's Bible in 40 years. Thanks to the invention of Gutenberg, and the egalitarian spirit of Martin Luther, religious communication flourished and crucial knowledge was made available and placed back into the hands of the community. Knowledge had been made accessible to everyone.

The fifth key element of COIN DNA is *internal honesty and transparency.* This is illustrated by the story of the Rothschilds.

Collaboration, knowledge sharing, trust, and reciprocal principles have been a recipe for sustainable success—not only in religion, but in business dealings back to the times when business was conducted by bartering. Forbidden to own land or join craft guilds, medieval Jews were confined to banking and finance. They had created a pan-European banking system based on reciprocal trust, in which Jewish bankers were able to draw on accounts of fellow Jews in foreign cities, knowing they would have to reciprocate the favor. This system worked perfectly, based on a mutually established code of ethics that forbade cheating. It reached its climax when, more than 200 years after Fugger, the Jewish banking dynasty of the Rothschilds created a banking empire that similarly dominated the European financial system for almost a century.

Born in 1744 in Frankfurt, Germany, Mayer Amschel Rothschild sent his five sons to the major European cities. Relying on trusted family members in all the European financial centers, the Rothschild bankers not only were able to transfer money quickly without having to transport it physically, but they also constructed an efficient information network based on carrier pigeons linking their different branches in major European cities. The second generation of the Rothschilds accumulated immense wealth in London, Paris, Vienna, Frankfurt, and Naples.

The Rothschilds followed their own ethical code of internal transparency and reciprocity. Each family member had to inform the others openly about his business dealings, while keeping non-Rothschilds out (only the male members of the Rothschild family were let into the innermost banking circle). For more than 100 years, the close-knit Rothschild banking empire dominated European banking, until the telegraph democratized access to information for everybody.

Over the course of history, these five elements have characterized teams of collaborative innovators, starting at the dawn of humanity. Disruptive innovations like gunpowder, the printing press, paper money, and bookkeeping are not the result of a carefully planned innovation process. Nevertheless, by analyzing teams of collaborative innovators of the past, we can identify their typical behavioral patterns. Those lessons from the past form five corollaries.

1. Collaboration networks are learning networks.
2. Collaboration networks need an ethical code.
3. Collaboration networks are based on trust and self-organization.
4. Collaboration networks make knowledge accessible to everybody.
5. Collaboration networks operate in internal honesty and transparency.

The elements constitute the "genetic code" of participants in COINs. And this genetic code of creativity, collaboration, and communication is deeply ingrained in the members of the cyberteam that make up a COIN. Linux developers offer us a modern example (see sidebar 3.1).

While the concept of innovation teams ranges back in history, the connective power of the Internet has greatly sped up the dissemination of new ideas, giving charismatic leaders with innovative schemes immediate global reach. But it is still the brilliant genius who comes up with the new idea, and

■

Sidebar 3.1
The Genetic Code of the Linux Debian COIN

Debian is the nonprofit distribution of Linux, competing with commercial distributions sold by Linux vendors such as Red Hat or SuSE. The Debian software developers are a vibrant community that has reinvented itself at least six times during the last eight years—exchanging leaders, changing how leaders are elected, administering its own social contract, creating the Debian constitution, and overhauling the voting procedure. In the process, it has become an increasingly efficient community.

The Debian community makes sure that its members share the right genes as highly skilled developers. Applicants first need to prove their worth by recommending changes. If those changes are accepted, a subsequent application must be sponsored by an accredited developer. Officially initiated members of the Debian COIN demonstrate five collaborative genes in an exemplary fashion. Motivational drivers like those introduced in the previous chapter—to satisfy one's ego, gain more knowledge, and have fun—blend seamlessly into five characteristics:

1. There is a *willingness to experiment.*
2. The Debian social contract is an explicit code of *ethics.*
3. COIN members follow the Golden Rule, which builds *mutual trust.*
4. An essential credo is to *make knowledge accessible to everybody.*
5. The entire COIN is based on *internal honesty and transparency.*

In fact, the Debian community goes much further than, say, the Rothschilds, committing itself not only to internal but also to *public* transparency. For instance, the Debian social contract explicitly states "We Won't Hide Problems," and all problems are published on a public Web site.

■

the group of collaborative leaders who get their team of disciples together to spread out the word—the Internet is only the catalyst.

Collaborators in Today's Knowledge Networks

Homo sapiens, Fugger, Renaissance Europe, the Reformation, and the Rothschilds exhibit characteristics of collaborative innovators working together in knowledge networks well before there was an Internet. They all worked as teams that embraced change, performing together in a collaborative way while conducting their business according to the ethical principles of their time. They shared knowledge internally, always ready to learn and to adapt to a changing environment. They share these traits with modern-day collaborative innovators such as open source programmers, the group that developed the World Wide Web, or the Internet Engineering Task Force (IETF) Internet standardization group.

There are some basic differences between the pre-Internet innovation networks and the Internet-enabled COINs of today. Members of innovation networks of the pre-Internet age had a fundamentally different attitude toward authority, transparency, and meritocracy. But the biggest difference is the existence of the Internet, which gives today's community members immediate global reach at very low cost; it allows teams to collaborate at a level of transparency never before possible. Unlike in earlier history, today's innovations do not trickle down through communication by carrier pigeons. COINs have reached their own tipping point, disseminating new ideas and concepts on a global scale—not in centuries, but in days and weeks. Linux and the World Wide Web became success stories within just a few years. Modern-day communication capabilities act as a huge catalyst, with the Internet giving the final boost. As instantaneous communication with anybody to anywhere is essentially free over the Internet, trust building, collaboration, and exchange of ideas can all happen at the speed of light. A second fundamental difference is that, whereas leaders of early innovation networks demanded unquestioning authority, the Internet-enabled COINs of today operate as true meritocracies. While Tim Berners-Lee and Linus Torvalds have reached guru status, members of the Web consortium and Linux developer communities obtain higher status through outstanding contributions and hard work transparent to all. In the Fugger and Rothschild banking empires, promotions were based on the decisions of family patriarchs.

The people making up today's COINs normally cross organizational boundaries. They form a dedicated community of members existing outside conventional organizations but sharing a common vision. Membership and roles within the community are based on contribution and merit, not on external hierarchical status. The community develops its own internal hierarchy based on skills and contributions, and people join the community not for immediate monetary reward, but because they believe in the goals of the community. Today's COIN members follow their leaders not because those leaders demand unconditional hierarchical authority, but because of shared beliefs, values, and goals. Thanks to the speed and open architecture of the Internet, today's innovation networks operate in internal transparency, and leaders are continuously challenged by their community. In the 21st century, the roles of COIN leaders and members are converging.

The essence of a community is its members. Membership in COINs is voluntary, rather than prescribed. Members are self-organizing and participate because they get high personal satisfaction from their participation; they are highly self-motivated and committed to their cause. Additionally, they are intellectually curious, are opposed to opaque hierarchies, and have a strong egalitarian mindset. Intel's Lablet COIN is a superb example (see sidebar 3.2).

But what about the communication aspects of COINs? What are the mechanisms by which people in COINs connect with others in their COIN and with the outside world? How are innovations disseminated?

Marco Polo: Collaborative Communicator of Innovations

Marco Polo may be the most famous world traveler of all time. His travel reports—written in a factual style that set him apart from his contemporaries[2]—became part of the mental climate of Europe. His writings shaped the perception of European geographers, adventurers, humanists, and Jesuits for centuries, as they set out to "discover" the world. The learned from all over Europe consulted him in Venice on aspects of academic geography. During his lifetime alone, Polo's book was translated into French, Franco-Italian, Tuscan, Venetian, Latin, and German—unprecedented for a book published in the Middle Ages.

Polo's travel reports stuck to the facts. He kept himself in the background, and he focused not on adventure but on providing a handbook for travelers and merchants—for example, giving the first detailed account to the West on the provenance of spices and other trade goods. This approach was immensely effective. As a collaborative communicator of innovations, he succeeded in "changing the world." As an archetypical communicator of innovation, Polo followed four simple rules and intuitively established four principles of open and honest communication:

Sidebar 3.2
Innovation Networks at Intel Research

Intel Advanced Research consists of a collection of COINs that communicate with each other in a small-world network. Under the leadership of David Tennenhouse, the company's director of research, Intel outsources a crucial part of its advanced research to its network of "Lablets"—small research labs, co-located at prominent universities such as the University of California at Berkeley, the University of Washington—Seattle, Carnegie Mellon University, and the University of Cambridge in England. Lablet staffs comprise an equal share of (a) professors and doctoral students on temporary leave from the university and (b) Intel employees. Although Tennenhouse and the Lablet directors define the overall strategic research agenda, researchers at the Lablets have wide-ranging freedom to pursue their own interests. Research teams, usually spanning different Lablets, form their own smaller COINs, while the combined Lablets form a knowledge network that reaches into the hosting universities and the central Intel research labs.

PlanetLab is one of Intel Advanced Research's early successes. A network of computers located at sites around the world, PlanetLab forms a platform for creating and deploying planetary-scale services—massive applications that span a significant part of the globe—such as global content distribution and file sharing. Applications running on PlanetLab are decentralized, with pieces running on many machines spread across the global Internet. They can also self-organize to form their own networks. No university could afford to deploy the large network of computers required to create PlanetLab, so Intel jump-started the project by contributing the first 100 nodes, the initial software implementation, and key sites. After some years of nurturing in the Intel Lablets, digital video content distribution has been announced by Intel and Hewlett-Packard as a first PlanetLab commercial application.

Intel Lablets leverage the latest results of high-tech research. Teams collaborate to achieve shared goals, while concurrently competing with and learning from their competitors. The rules are transparent and clear for everybody, and the teams function as a meritocracy.

- *Stick to the facts.* Polo described only what he saw himself. He sometimes might have had difficulty placing some things that he had seen in context, but by and large his assertions were proven right by later voyagers.
- *Make the message objective.* Unlike some other travel writers of his time, Polo objectively portrayed the lifestyles of those with whom he came in contact. He held no grudge against the Mongols, for instance. His descriptions included truly useful tips for his fellow merchants and voyagers.
- *Walk the talk.* Marco Polo really did spend 24 years with the Mongols. His 17 years at the court of the Great Khan gave him unique insights into how the Mongolian empire was ruled, which he shared with his readers.
- *Stay in the background.* Polo focused on the places he went and people he met.

These principles have not lost their appeal over the last 750 years. To the contrary, they have never been more true than they are today—thanks in large part to the transparency brought to us by the Internet.

Collaboration on the Internet Is Transparent

When other early Web enthusiasts joined Tim Berners-Lee and Robert Cailliau, no one was officially in charge. Individuals decided what they wanted to work on—whether they wanted to set up a Web site, develop a new browser, or implement another Web server. They volunteered for the roles of Web master, project manager, software developer, or user interface designer based on perceived project needs and their own strengths and weaknesses. No one joined the Web project to make money, get a promotion, or become famous; they joined because they were convinced of the merits of the Web and wanted to make the vision of Vannevar Bush and Ted Nelson come true. They had an intrinsic motivation to work toward a common goal.

Everyone joined the project team of his or her own free will. No one was ordered to do so by superiors or managers. To the contrary, since the goals of the Web initially were of only tangential interest to the organizations team members worked for, people had to fight for time and resources to be able to join this COIN.

What made efficient collaboration over long distances possible was that all the knowledge was shared openly among COIN members. One of Berners-Lee's most crucial decisions was to base the Web, from the beginning, on public standards such as Standardized Generalized Markup Language (SGML). He immediately published his document specification standard, which he called HTML (Hypertext Markup Language), and the source code of all the software he developed. The people who joined him did likewise, distributing all early Web server and browser software under the open source software distribution license. They also met face to face whenever possible. And all their work in progress about the Web was, from the outset, published on the Web. This meant the people working together on developing the Web operated in an environment of transparency and honesty.

Communication on the Internet Is Honest

Researchers have discovered the Internet encourages honest communication (which makes it easier for COIN members to tell each other the truth). For example, Cornell professor Jeffrey Hancock asked 30 of his undergraduates to record all of their communications and all of their lies over the course of a week.[3] Evaluating the results, he found that the students had mishandled the truth in about one-quarter of all face-to-face conversations, and in a whopping 37 percent of phone calls. But in cyberspace, the students became much more honest: only one in five instant-messaging chats contained a lie, and barely 14 percent of e-mail messages were dishonest.

It is the fear of exposure that makes us more truthful online. In "real" life, after all, it is actually quite easy to get away with a little white lie now and then. If you embellish the truth to someone at a cocktail party or on the phone, you can always backtrack later and claim you said no such thing. In all likelihood, no one recorded your conversation. On the Internet, though, your words often come back to haunt you.

The digital age is tough on its liars, as a seemingly endless parade of corporate executives is learning to their chagrin. Today's industry moguls are trapped not by ruthless competitors but by prosecutors collecting transcripts of old e-mails, filled with suggestions of deception. Even Microsoft was caught off-handedly by old e-mail messages in its defense against antitrust lawsuits. But this is not a problem only for corporate barons. We all, in principle, know that in cyberspace our words never die, because machines don't forget—an idea built into the very format of e-mail when programs automatically "quote" your words when someone replies to your message. Still, it is not only the fear of electronic exposure that drives us to tell the truth. There is something about the Internet that encourages us to spill our guts, often in rather outrageous ways. Psychologists have noticed for years that going online seems to have a catalytic effect on people's personalities. The most quiet and reserved people may become deranged loudmouths when they sit behind the keyboard, staying up until dawn and conducting angry debates on discussion boards with total strangers.

One group of psychologists found that heated arguments—so-called flame-war fights—were far more common in online discussion boards than in comparable face-to-face communications. Open University U.K. psychologist Adam Joinson conducted an experiment in which his subjects chatted online and in the real world. He found that, in online communication, people are more likely to offer personal details about themselves without any prompting.[4] Joinson also noted that the Samaritans, a British crisis-line organization, found that 50 percent of those who write in via e-mail express suicidal feelings, compared with only 20 percent of those who call on the telephone—not because Internet users are more suicidally depressed than people offline, but because they are more comfortable talking about their feelings in the anonymous privacy of their own rooms.

Our impulse to confess via cyberspace inverts much of what we think about honesty. It used to be that if you wanted to know someone—to really know and trust them—you arranged a face-to-face meeting. Our culture still emphasizes physical contact, the shaking of hands, and the small talk. Executives and politicians spend hours flying across the country for a five-minute meeting on the assumption that even a few seconds of face time can cut through the evasions of letters and legal contracts. But, as more and more of our daily lives move online, we might find ourselves living in an increasingly honest world, or at least in one where lies have ever more serious consequences. With its persistent long-term memory, the Internet gets us to stick to the truth.

COIN members, mostly collaborating over the Internet, have accepted and adapted this insight. Groups such as the Linux Debian developer community even vow publicly in their constitution not to hide any defects and problems arising from their development work.

From everything we've discussed thus far, one would think that COIN members must have the latest in Internet technology to communicate with each other—but that is not the case. The online campaign of a friend of mine for a seat in the Kenyan parliament shows that bare-bones e-mail accessed from Internet cafés is entirely sufficient.

Low-Tech with High Impact: Campaigning for Kenya's Parliament

I first met Tom Ojanga when he was a student in Germany and was looking for a summer job in the information technology (IT) department of Union Bank of Switzerland (UBS). He had come to Germany from Kenya at age 19 and won a scholarship to study computer science in Konstanz. After completing his studies, he became a software engineer at UBS in Zurich, quickly climbing the promotion ladder to the rank of assistant vice president. After six years with the bank, Ojanga joined Deloitte Consulting as a manager.

Throughout his professional career, Ojanga worked to improve conditions in his native country. While a student, he became president of the African Students Association in Germany. He also helped found the Kenyan Association for the Advancement of Computer Technology (KaACT), which paved the way to Internet connectivity in sub-Saharan Africa. The original members of KaACT founded Africa's leading Internet service providers, AfricaOnline and ARCC. After school, Ojanga became chairman of the Worldwide Kenyan Congress, an organization that brings together other Kenyan organizations in the Diaspora. This was the beginning of his forays into a political career.

Ojanga decided to quit his position at Deloitte Consulting in 2002 and run for a parliamentary seat in Kenya. Although he had not lived in Kenya for 16 years, "The Tom Ojanga Initiative" was well known among the many poor who comprise Kenya's lower strata of society—a group of people with whom the ruling elite physically located in Kenya can rarely connect meaningfully. Ojanga communicated regularly with women and youth from several Kenyan villages, and answered to the needs of thousands of people. People asked him daily for help obtaining supplies, medicine, or any of a host of necessities that are rarely available to the poor in Kenya's impoverished economy.

Considering the enormous geographical distance between Kenya and Ojanga's place of residence in Switzerland, how did he manage daily communication with rural and isolated parts of this East African nation? The answer may be surprising. He instituted a virtual COIN using simple Internet technology to connect himself with women and children in Kenya, as well as with politicians and businesspeople around the world. This network shares one common goal: the eradication of the daunting disparity between rich and poor in Kenya.

The birth of this network lies in an organization called Swissimpact, which houses the "Computer for Schools" project Ojanga founded while still at UBS. Swissimpact ships hundreds of old, but still usable, computers donated by the bank and others to Kenya. In 2001, Ojanga again presented used computers, which were donated by the Swiss Deloitte Consulting practice, to Kenyan schools This provided an Internet connection to 46 rural schools (whose pupils were mainly disadvantaged children), fostering knowledge sharing and communication with European schools. The establishment of collaboration channels between local and other schools abroad was covered widely in the local media and attracted a lot of attention from other parts of society; from then on, Ojanga would receive requests for help from women's groups, street children, youth groups, and others.

To set up his COIN, Ojanga made contact with literate, and sometimes semiliterate, persons who could represent a group ranging from 300 to 3,000 individuals. He then helped these contacts create free e-mail accounts through portals such as Microsoft's Hotmail and AfricaOnline's e-Touch. He maintained regular weekly communication with women's groups, youth groups, and disabled groups who organized themselves to provide answers to their needs; group leaders would print Ojanga's messages, which would then be read to group members. The messages were often announced by the group leader, who walked through the village ringing a bell. Similarly, messages were relayed back to Switzerland through Ojanga's Internet café contacts. The $1 cost of printing a message and sending replies was handled by his campaign at the end of each month.

Ojanga's Poverty Liberation Project was a key element of his campaign. Its purpose was to empower students; orphaned and neglected children; battered and disadvantaged women; and disabled people suffering from drug, sexual, and labor abuse, malnutrition, and disease. Ojanga's COIN has been a highly successful part of this campaign. It has beaten the odds of collaboration across overwhelming geographical, cultural, and economic boundaries by connecting global sponsors and businesses with various Kenyan nonprofit organizations. Ojanga's initiative helped obtain needed supplies, books, clothing, and housing from willing donors around the globe and deliver them to these organizations. In addition, Ojanga has been able to use his political influence to prevent squatters from being kicked out of their homes, to convince sponsors to provide the medicine and hospital volunteers necessary to administer typhoid vaccinations, and to lobby for the rights of women.

For Ojanga, the reward was not only helping his native people; it was also in the creation of a real persona and presence among the large poor population in Kenya, despite his physical absence. Naturally, this helped his campaign tremendously. This kind of contact is rare in Kenya, and many Kenyan elites do not approve of publicizing their domestic social and economic problems internationally because it threatens their power. When Ojanga visited Kenya during his campaign, he was even jailed by the then ruling party on fabricated accusations, although he was released after one night.

Ojanga's political campaign illustrates the power of the Internet to connect a global COIN, at the same time bridging continents and the digital divide—without needing the latest technology, but employing bare-bones e-mail to connect with barely literate single mothers in rural Kenya.

While his example illustrates that e-mail can be sufficient to operate a global COIN, Microsoft is recommending the latest Internet and computer technology to build its digital nervous system.

Microsoft—Building the Digital Nervous System

Microsoft exhibits many of the traits of an organization supportive of COINs. Bill Gates, in the introduction to his book *Business @ the Speed of Thought,* describes the core tenets of COINs:

> Insist that communication flow through the organization over e-mail so that you can act on news with reflexlike speed. . . .
> Use digital tools to create cross-departmental virtual teams that share knowledge and build on each other's ideas in real time, worldwide. . . .

> I believe in a very open policy on information availability. . . . The value of having everybody get the complete picture and trusting each person with it far outweighs the risks involved.[5]

Microsoft has built up a culture of reacting quickly to change. The company's leader is similar to a COIN leader, having obtained guru status within Microsoft. For example, the company dealt marvelously with the emergence of the Web. Gates immediately recognized the threat and disruptive potential of this new technology to Microsoft's monopoly on desktop operating systems, and he swiftly turned his company around, firmly embracing the Web browser concept. Microsoft quickly bought and integrated a broad collection of Web technology companies, overtaking Netscape as the leading Web browser provider and turning a formidable challenge into a major victory.

Microsoft has built its innovation process on two pillars: first, it supports transformative innovation internally; second, it acquires disruptive innovations externally by skillful application of the open innovation process. Microsoft has been quick to embrace the open innovation paradigm. Gates announced that, for fiscal year 2004, Microsoft would extend its already huge research and development budget by an additional $500 million, reaching a total of $6.8 billion. Most of this money was invested internally in transformative innovations—for example, improving speech recognition based on speech-recognition technology the company already owns. Microsoft is also extremely efficient leveraging the disruptive innovations of others by acquiring innovations—buying startup companies or hiring innovative researchers. Examples of this abound throughout Microsoft's history. To name just a few, the company purchased DOS from another small Seattle company, Seattle Computer; the Windows user interface is similar to the Macintosh interface; Internet Explorer is similar to Netscape Navigator; and Excel is an improvement on the VisiCalc spreadsheet application, which became available on IBM personal computers in 1981.[6]

The second distinctive feature of the Microsoft innovation process is its selective sharing of knowledge with others. On the one hand, as the dominant representative of closed source software development, Microsoft keeps the sources of its software tightly controlled. On the other hand, starting with DOS and Windows, Microsoft has always published the hooks into how to program its applications, and it has encouraged third-party developers to develop software for its operating system and its stack of applications. This

Sidebar 3.3
The Problem with Software Patents

The patent system was created at a time when innovation consisted primarily of new hardware. No one really imagined then that someday pure ideas would also be patentable. In fact, before 1980, patents were rarely awarded for business methods or computer software. The courts deemed a software program to be a mathematical algorithm—that is, simply a language that tells a computer what to do. In the 1980s, though, courts began to rule that software, if it produced something useful, merited patent protection.

Thus, originally created to stimulate innovation, patents may now be having the opposite effect—at least in the software industry.

Abuse of software patents has become widespread. For instance, an inventor who held a patent on a widely used feature of Web browsers kept the existence of his patent unknown for years and then sued Microsoft for patent violation. The court awarded him $500 million. While this decision is currently being appealed, the U.S. Patent Office is reviewing the validity of his patent. In another example, British Telecom claims to own the patent that covers the *idea* of hyperlinking, a concept used nowadays by everyone who surfs the Web.

One of the main reasons companies apply for large numbers of software patents is to build up their defensive portfolio of patents in case they need to countersue when someone makes a claim. But it appears that large companies would prefer to avoid expensive and uncertain litigation rather than assert their own intellectual property rights, and most large software companies enter into cross-licensing agreements in which they agree not to sue over patent infringements.

Even IBM has adopted a critical view on software patents. In early 2005, IBM made 500 software patents freely available to developers and users of open source software. The goal is to stimulate innovation and interoperability through open standards. Says Adam Jollans, IBM's worldwide Linux strategy manager: "This is about encouraging collaboration and following a model much like academia."

■

Sidebar 3.4
Silicon Valley and Route 128 Illustrate
the Commercial Advantages of COINs

Route 128, which rings the Boston metropolitan area, is called "America's Technology Highway" because it is where many high-tech companies have gotten their start. Everyone knows of California's Silicon Valley, which eclipsed the Boston area as the high-tech Mecca during the Internet boom of the 1990s—but Boston is rebounding.

Researcher Annalee Saxenian identifies some huge commercial advantages in the incubating business environments of these two areas. Although she doesn't refer to them by name, she is in fact talking about COINs: "The contrasting experiences of Silicon Valley and Route 128 suggest that industrial systems built on regional networks are more flexible and technologically dynamic than those in which experimentation and learning are confined to individual firms."[1]

Even Saxenian's main explanation for the success of Silicon Valley companies over Boston area enterprises speaks to the power of COINs. The California firms have shown a more open environment and a culture of trust and collaboration. While employees at the greater Boston corporations have tended to look for innovation and resources within their own organizations, executives of Silicon Valley firms have shown a far greater willingness to collaborate over company boundaries.

---

is quite different from Apple Computer's past behavior, which in the first 10 years of its existence "built a fortress to protect themselves, but found out they are isolated from the rest of the industry."[7]

Quite understandably, Microsoft also demonstrates some non-COIN-like characteristics—for example, relentlessly fighting open source software. Microsoft's response to the threat of open source Linux to its desktop operating system monopoly has been to redouble its efforts to persuade customers that its own software is worth paying for and to fine-tune its marketing and development machinery to make sure it keeps its customers happy. It has also started to issue warnings that the open source approach discourages innovation.

So far, though, Microsoft has not been very successful in this crusade. Rather than displaying lack of innovation, the open source community has produced a long list of highly innovative software programs of high quality,

The East Coast firms are learning their lesson and breaking down corporate boundaries for the sake of more flexible organizational forms. For instance, research universities in the greater Boston area are playing a crucial role as the area's special economic advantage in the form of magnets for talent and investment—infusing more than $7 billion into the regional economy each year.[2] Boston forms a geographically co-located innovation network; it has the world's highest density of universities, Nobel Prize winners, startups, and venture capital firms.

Collaboration networks of academic researchers have a culture of COINs. Academic research environments, while rigidly hierarchical, are also *meritocracies* based on peer review and reputation. *Transparency* is valued highly; pressure to "publish or perish" means that ongoing research projects at universities are highly visible to the outside world, and researchers routinely share information at conferences, seminars, and workshops. Finally, because universities are tremendously aware of research ethics, there is a high degree of *consistency*.

In a "small world" like the Boston academic institutions, researchers act as hubs of innovation and trust, linking a highly productive community of innovative knowledge workers.

1. Saxenian, *Regional Advantage: Culture and Competition in Silicon Valley and Route 128* (1994).
2. For more information, see *The Harvard Gazette* at
   www.news.harvard.edu/gazette/2003/03.13/01-economic.html.

ranging from the Apache Web server, Linux, and the Mozilla Firefox browser to free desktop publishing software. Nevertheless, until now, Microsoft has been masterfully exploiting the strength of its internal innovation networks, while at the same time connecting with the fruits of COINs from outside the company through the open innovation process.

As the main proponent of a digital nervous system, Microsoft has recognized that communication technologies today have reached their tipping point, where asynchronous technologies, such as e-mail, and synchronous communication by phone, chat, and Web conferencing converge in what I call "many-to-many multicast." (These changes are discussed in detail in chapter 6.)

Unlike Microsoft, Swiss House for Advanced Research and Education (SHARE)—the first Swiss digital consulate in Boston—does everything it can to connect innovators physically and virtually in an open environment.

SHARE—Linking Swiss Entrepreneurs
into the Boston Innovation Network

Sitting at the large communal table in the main lobby of the SHARE building located in Cambridge, Massachusetts, between the Massachusetts Institute of Technology (MIT) and Harvard, Christoph Von Arb, Swiss consul general and director of SHARE, and Pascal Marmier, responsible for innovation and entrepreneurship (who has been with SHARE from its beginning), explained to me SHARE's basic principle. SHARE defines itself as the world's first digital consulate, linking the scientific, academic, and high-tech communities of New England and Switzerland. SHARE is a community that is both physical and virtual. It serves as a face-to-face meeting point for creative thinkers and entrepreneurs from both areas of the world. An interactive Web site with video feeds from SHARE allows people to tune into live happenings at SHARE and to access archives of past events. Together with second digital Swiss consulate Swissnex, located in the Silicon Valley, SHARE provides an entry point for Swiss entrepreneurs and researchers into the high-tech worlds of New England and California.

SHARE extends local collaboration incubators into a global innovation network by forming a bridge linking innovators abroad and in the greater Boston area.[8] SHARE works for example, with MIT's Deshpande Center, which fulfills its mission to "bridge the innovation gap" by connecting academic researchers, young entrepreneurs, and the outside business community through symposia, education, and other efforts.

Whether in a globally active research and development community, a corporate setting, or in a scientific incubator environment like the Boston area, meritocracy, transparency, and consistency define a milieu in which COINs thrive. By providing a supportive culture, organizations can enjoy the fruits of their COINs' work.

There have always been organizations that exhibit those traits. However, only in the Internet age does the COIN concept gain its full advantage, where these three crucial traits—meritocracy, transparency, and consistency—are augmented by the many-to-many multicast capabilities of the Internet. The transparency of Web-based communications bolsters instantaneous collaborative networking on a global scale. By linking communities of innovation, organizations such as SHARE and MIT's Deshpande Center have emerged as exemplary communication hubs for COINs.

The DNA of COINs that we've learned about in this chapter is augmented by two other critical elements: the ethical code by which COINs operate, and COINs' small-world networking structure. Understanding these elements—the subject of our next chapter—will allow us to grasp the value to businesses of COINs, illustrated by the real-life examples in chapter 5.

CHAPTER 4

Ethical Codes in Small Worlds

Collaborative innovation networks (COINs) are self-organizing, unified by a shared vision, shared goals, and a shared value system. COIN members communicate with each other in a "small-world" networking structure where each team member can be reached quickly.

How does this self-organization really work? COIN members are brought together by mutual respect and a strong set of shared beliefs. These common values act as a substitute for conventional management hierarchies, directing what every COIN member "has to do." COINs have internal rules by which they operate, for how members treat each other, for how supportive behavior is rewarded, and for how members are punished when they do not adhere to the code. There is a delicate internal balance of reciprocity, and a normally unwritten code of ethics with which members of the COIN comply.

Ethics are a central issue in today's business world. The headlines about the breakdown of ethics at WorldCom, Tyco, Enron, and elsewhere have forced senior managers worldwide to take a harder look at how their companies "behave" in the world. COINs help make sure an organization behaves "right"—and help keep an organization innovating in its core business, *not* in creative accounting.

How People Attach to COINs

COINs share many similarities with other types of egalitarian communes that may seem anathema to the business world (such as the hippie communities of the 1960s, or some fundamentalist religious communities). The shared vision in COINs or the ideology in communes is the foundation upon which the common understanding of the community is built. COINs share the commitment mechanisms of a commune—they foster a strong sense of unity through communal sharing, labor, regular group contact, and homogeneity. As with communes, COIN members might even experience some form of "persecution"—for example, because they decide to allocate some of their company working time to the goals of the COIN rather than to the tasks given them by their superiors. And, also similar to communes, the commitment of COIN members increases if they are not immediately compensated but instead see their work as an investment in a better future for themselves and their environment.

Rosabeth Moss Kanter, a professor at Harvard University, has identified six commitment mechanisms of communes that explain why members become detached from their previous social environments and attached to their new communes (see table 4.1). These six processes also apply to building commitment in COINs. In particular, the commitment-building processes "investment" and "communion" are crucial to the success of high-functioning COINs.

Table 4.1. Commitment-Building Processes of Communes.

Detach	Attach
Sacrifice Individuals give up something valuable to belong to the community.	*Investment* Individuals gain a stake that they will loose if they leave the community.
Renunciation Individuals give up competing relationships outside of the community.	*Communion* Individuals acquire a "we-feeling" by belonging to the community.
Mortification Individuals give up their former identity to gain a new community-controlled identity.	*Transcendence* Individuals will base part of their decisions on the will of the community

Data from Kanter, *Commitment and Community* (1972).

- *Investment.* Over time and by active participation, individuals will gain a stake in the group, so they must continue to participate if they are to realize a profit on their investment. For example, in many communes, members gain a more privileged status in the community with increasing seniority. In a COIN, members who have invested time and effort to learn the language, rules, and skills to become respected core team members will not easily relinquish this elevated status within their community.
- *Communion.* Individuals build relationships inside the community and get into meaningful contact with the collective whole, so "they experience the fact of oneness with the group and develop a we-feeling."[1] For example, some communes have regular sessions where members are expected to confess their "sins" to the community and obtain absolution and advice on how to improve their personalities. COINs create a virtual feeling of communion and build relationships by working together toward the shared goal by exchanging e-mails, collaborating in online group chat rooms, meeting face-to-face and virtually, and using a shared language.
- *Transcendence.* Individuals attach their decision making to a power greater than themselves, surrendering to the *group's* higher meaning and submitting to something beyond the individual. In COINs, once members have bought into the common vision and goals, they become extremely committed to working toward reaching those goals, with little regard for outside influences. They exhibit self-organizing swarming behavior whose purpose is not always immediately obvious to those watching from the outside. (Moss Kanter's detachment mechanisms do not apply to COINs as strongly as they do in other situations. However, COIN members can experience "weak symptoms of detachment" from their outside lives.)
- *Sacrifice.* Individuals give up something considered valuable or pleasurable to belong to the commune. In the context of a COIN, members might forgo other short-term external rewards when they decide to work for the vision and goals of their COIN.
- *Renunciation.* Individuals give up relationships outside the community. COINs do not request such renunciation by members,

but highly committed core team members might spend so much time working in their COIN that they neglect their outside relationships.

• *Mortification.* Individuals exchange their former identity into one defined and formulated by the community. In COINs, members are respected for the skills they possess and their achievements relevant to the community, with little respect for their outside position. But while COIN members in a virtual community of programmers might be known by their e-mail names and have a virtual personality reduced to their programming capabilities, they are definitively not requested to give up their outside identities.

The Internet Engineering Task Force (IETF) working groups, which developed the basic standards of the Internet, illustrate how the commitment-building mechanisms investment, communion, and transcendence are employed by successful COINs. IETF members are expected to display an egalitarian work ethic similar to that found in communes:

IETF . . . working groups go through phases. In the initial phase (say, the first two meetings), all ideas are welcome. The idea is to gather all the possible solutions together for consideration. In the development phase, a solution is chosen and developed. Trying to reopen issues that were decided more than a couple of meetings back is considered bad form. The final phase (the last two meetings) is where the "spit and polish" are applied to the architected solution. This is not the time to suggest architectural changes or open design issues already resolved. It's a bad idea to wait until the last minute to speak out if a problem is discovered.[2]

However, IETF members certainly are not expected to detach from their "previous" or "outside" life. They might become attached to their IETF working groups, undergoing the processes of investment, communion, and transcendence by adhering to the "Tao of IETF" described below.

Much like communes, COINs tend to set themselves apart from the rest of the world by using a language and symbols only understood by themselves. Defining this common behavioral code includes different components. Once a COIN has been established, the brand image will attract like-minded people.

The Tao of COINs

> The Internet has become a grossly commercialized Wild
> West in so many ways. But the community spirit on which it
> was founded is alive and well. The Net depends on the same
> spirit that motivates volunteers in the physical world: a
> commitment to solve problems and make life better for those
> who might otherwise not have the resources or expertise.
>   . . . [T]here are thousands . . . who quietly do their best
> for the larger community. They run e-mail lists and maintain
> software archives, fight viruses and bugs, and so much more.
> They maintain an old-fashioned credo of altruism in an era
> when the idea of a commons is under attack.[3]

Members are attracted to a COIN by sharing a common culture of honest communication and transparent collaboration. (These are the two factors that clearly distinguish COINs from religious communes.) COIN members exhibit a behavioral pattern defined by high interactivity, high connectivity, and a high degree of knowledge sharing. These behavioral patterns and cultural characteristics lead to a code of ethics for COINs.

An ethical code sets down the rules and principles that should be followed by all associates of a group. Because COIN members are deeply and intrinsically motivated citizens of their community, they stick to their code of ethics in much the same way that citizens of a country stick to their laws and following the rules of their society. Unlike in countries, though, where positive behavior usually goes unrewarded while offensive behavior is punishable by law, the unofficial reciprocal ethical code *does* reward COIN members for positive behavior.

For COINs, we can derive an ethical axiom from the "theory of justice" of philosopher and former Harvard law professor John Rawls:

> All social primary goods—liberty and opportunity, income and wealth,
> and the bases of self-respect—are to be distributed equally unless an
> unequal distribution of any or all of these goods is to the advantage of
> the least favored.[4]

In a COIN, the primary good to be distributed is knowledge, and the most knowledge is to be given to the least knowledgeable members of the COIN.

This is to the long-term benefit of the COIN, because the more knowledge individual COIN members gain, the more productively they can work together toward their common goals. Of course, junior members are expected to be active gatherers of knowledge, respectfully contacting the experts. In return, knowledge experts are supposed to make themselves available to answer the questions of the novices.

There is a delicate internal balance of reciprocity, and a normally unwritten code of ethics that is adhered to by members of a COIN—the "Tao of COINs," from the Chinese word meaning "The Way" or "The Law" or "The Rule."[5] Various Chinese philosophers, writing in the 4th and 5th centuries B.C., presented major philosophical ideas and a way of life that are now referred to as Taoism—the way of correspondence between man and the tendency or the course of natural world. In Taoism, development of virtue is one's chief task. The three Tao virtues to be sought are compassion, moderation, and humility—all qualities crucial for the success of COINs.

The Tao of the IETF gives a practical example of an ethical code for a virtual community of innovation. It defines the rules, how the IETF operates, and how its members work together.

The purpose of this . . . is to explain to the newcomers how the IETF works. This will give them a warm, fuzzy feeling and enable them to make the meeting more productive for everyone. This . . . will also provide the mundane bits of information which everyone who attends an IETF meeting should know.[6]

In the IETF, the mailing lists of the working groups are the main means of communication. Active IETF members usually meet just three times per year. The Tao of IETF defines the etiquette and the ethical code for how IETF members treat each other online and when meeting face-to-face. It defines obvious things such as how to register for the meetings, and how to participate in the standards development process; it also contains the implicit rules of group behavior, such as the dress code, the social events taking place at the meetings, and the meaning of the colored dots on the badges of meeting participants.

The ethical code of a COIN is the main "glue" that holds it together, setting out the informal rules and principles that should be followed by all members of the group. Rewards in a COIN are given primarily in the form of peer recognition and admission to the core group, and punishment is by withholding recognition (or exclusion for really grievous offenders). The behavioral code of conduct in online communities can be traced back to the Golden Rule: do to others only what you would like others to do to you. From a reading of many IETF and other online community mailing lists, four rules emerge:

1. *Respect your elders.* While COINs have an egalitarian culture, COIN leaders or gurus define a COIN's future direction. Elders are respected not because of their hierarchical position, but because of their vision. They are also often among the most experienced subject matter experts of a community.

2. *Be courteous with your fellow members.* Members of a COIN are expected to treat each other with mutual respect. For example, "flaming" other COIN members in public by sending negative comments to a mailing list is a serious breach of etiquette. Rather, it is expected that negative comments be made in private in a constructive way.

3. *Say something only if you have something to say.* It is expected that junior members of a community acquire their knowledge not by asking "naïve" questions in public, but by studying the FAQ (frequently asked questions) lists and by privately consulting recognized knowledge experts. New members of a community are also expected to become knowledgeable as quickly as possible.

4. *Be ready to help your fellow community members.* Senior members, knowledge experts, and gurus are usually quite accessible. Recognized knowledge experts are expected to share what they know freely, educating more junior members so they become knowledgeable themselves.

The guiding principles of the Tao of COINs allow an intelligent questioning of the rules. COIN members demonstrate a "feel" for their community, treating each other with dignity and esteem. Similarly, leaders of COINs interfere relatively little with the daily activities of their community, letting it operate based on the shared code of ethics. While it is very clear that Linus Torvalds is in charge of directing the future development of Linux, for example, he does so in a very subtle way without using his hierarchical position and does not involve himself in the "fussy" details of day-to-day operations.[7]

■

Sidebar 4.1
An Illustration of Rules Governing How
COIN Members Treat Each Other

SOCNET is a group of sociology researchers interested in social networks analysis.[1] A recent discussion on the group's mailing list illustrates how four rules govern the way in which members of the SOCNET COIN treat each other.

Somebody asked a "nave" question, breaking rule #3 and wasting the bandwidth of the group. It provoked this reply: "It is amazing how smart people on this list do not know how to use Google!"

But because this reply was considered too rude and thus broke rule #2 by being too close to "flaming," the person who sent the reply was himself reprimanded, not only for breaking rule #2 but also #4 by not wanting to help a fellow community member: "Nothing personal here. Actually, from the past e-mails, I knew that you are one of the most active and willing-to-help persons in this list. But you know, for a newbie, a right pointer to the right article as start-learning-point can help them a lot and save them a lot of time."

In the discussion that followed, the original "nave" questioner was finally advised to apply rule #1 and show respect to his elders when asking questions: "It's been my experience on USENET and other forums where you are making a written (e-mail) request to a group of experts, that it's a good idea to show the work that you have already done to try and solve your problem or answer your question."

1. See www.sfu.ca/~insna/INSNA/socnet.html.

■

The Tao of COINs coordinates the working behavior of COIN affiliates. Usually COIN members carry this code "in their genes"—that is, they live by it without it being written out. Behaving according to these mostly unwritten ethical rules, COIN members are implicitly more sociable than members of many societies that live by a written rigorous law. As COIN members participate of their own free will and usually are not paid to work, they expect and exhibit fairness, compassion, and altruism. COIN members go out and help others because they expect others to do the same for them.

As the term implies, the Tao of COINs subsumes altruistic behavior, compassion for fellow COIN members, and moderation and humility of their leaders. Three fundamental terms of a culture for COINs summarizes the Tao of COINs—meritocracy, transparency, and consistency.

Meritocracy as an extension of reciprocity, the principle of fair treatment of everyone, is at the core of successful innovation communities. COIN members are willing to help and share what they know with others, but they expect similar behavior from other COIN members. Meritocracy is a direct consequence of the application of the Golden Rule. For example, open source software developers contribute their code because they expect their codevelopers to do the same so that all may enjoy the benefits of using a common software code base. The community judges the quality of each individual contribution, since each contribution is open to the entire COIN.

Transparency means that rules are made explicit, and the roles and responsibilities of every COIN member are obvious to the entire community. In a COIN, strengths and weaknesses of every member are exposed, but contributions are also made transparent. For example, the skills and the role of every programmer are obvious to all members working on a common open source software project. Each team member expects to get fair credit for her or his contributions; peer recognition is one of the main motivations for COIN participants.

Consistency means that every COIN member behaves according to the Tao of COINs, and delivers on promises made to the community. Under the same conditions, similar actions will produce the same results. Consistency is the basic requisite for fair treatment of all COIN members. For example, open source developers are expected to stick to the programming rules and guidelines that are in effect for their project. Members of the Debian Linux distribution have defined the Debian Social Contract to ensure consistent behavior of all COIN members, including their leaders.

Meritocracy, transparency, and consistency also mean that actions within the community are grounded in reason and not in randomness. Innovation communities are driven by learning, logic, and a shared vision of working toward "furthering the state of the art." In an open source software project, it is expected roles will be filled strictly on the basis of merit, and not because of previous relationships, hierarchical positions, or other criteria not relevant to the project. COIN members behave along those guidelines without being aware of doing so. The Tao of COINs—defined by meritocracy, transparency, and consistency—is part of their "genetic code" in the same way social insects exhibit swarming behavior, enabling them to work toward a larger goal as a

---

Sidebar 4.2
You Might Think They're COINs, but . . .

Let's look at organizations that are *almost* COINs. They may have the "innovator's gene" and operate with swarm intelligence, and they may have built virtual communities that rely on the Internet for internal communication. But without meritocracy, consistency, and transparency—essential for success—they aren't really COINs.

One such organization is the U.S. National Aeronautics and Space Administration (NASA), created to innovate. But its innovative capabilities are severely hampered by its rigid "command and control" structures. Open and honest communication is buried in NASA's huge hierarchy. The *New York Times* posits that the *Columbia* space shuttle disaster would not have happened if NASA had a more collaborative, meritocratic, and transparent culture.[1]

In a strong indictment of NASA management, crash investigators reported that the agency's work culture discouraged dissenting views on safety issues. The missing flow of information and the agency's approach to peer review and accountability prohibited its own engineers from following their instincts. For some of the deep technical analysis needed to make space flight successful, NASA was relying on external contractors who were afraid to speak out in clear language (an effect these contractors call "NASA chicken syndrome"). These are characteristics totally adverse to COINs.

The worldwide antiglobalization movement offers another illustration. The nongovernmental organizations (NGOs) and "affinity groups" that protest

---

self-organizing team, where each team member does not have to know the big picture in detail all the time.

Occasionally, organizations may appear to be COINs, or like COINs, because they display some of the typical characteristics. But the deficiencies can make all the difference, as sidebar 4.2 explains. When organizations lack meritocracy, consistency, and transparency, it is an indication of another serious deficiency—that of *social capital*, which is for society what financial capital is for the economy. Investment in social capital—reciprocity in social networks—is the key building block for a COIN culture. And an organization founded on meritocracy, consistency, and transparency accumulates social capital.

globalization, organizing demonstrations with huge turnout at events such as the G8 summits, operate much like COINs. Participants are members of a huge shared-interests network, unified by their disdain for our current industrial system of a globally linked economy, and in their rejection of the existing system of world governance. A smaller group of more active "practitioners" who organize the protests form a hard core. The movement's nucleus forms a COIN-like structure, with people sharing a common vision and cooperating actively toward their shared goal.

Antiglobalization protesters have much similarity to COINs. They run a creative organization, making highly efficient use of the Web to coordinate on a global scale with high agility. But they operate shrouded in secrecy, lacking the transparency that is a crucial success criterion for high-functioning COINs. This lack of transparency also masks that the movement is short on alternatives to the existing world order. The laudable *vision* of the movement to break down existing wealth inequalities is not coupled with many specific goals on how to bridge the divide—which means that the goals are not known and shared, as in a COIN. If the movement wanted to operate successfully as a COIN, it would have to develop more convincing goals, while at the same time laying open its ethical code, governance and leadership structure.

Finally, there are religious cults, which have strong similarities to COINs. They share self-organization and swarm intelligence, and they have the egalitarian properties of communes. But they lack the rationality and transparency essential to high-functioning COINs.

1. Leary, "Better Communication Is NASA's Next Frontier" (2004).

Social Capital Is the Currency of COINs

I have a friend who is a craftsman and who also breeds and races horses. Horse racing is a very expensive hobby. My friend is not particularly rich in financial capital, but he is rich in social capital: he is embedded in a rich web of social relationships and of trust, of giving and taking. He uses his skills as a craftsman and horse owner to do favors and build relationships with others—for example, by helping fixing little things in the household, or by allowing children of friends to visit his horse stable. In return, his horse trailer is serviced for free, and he gets materials for rebuilding a stable at a very low price. He is not trading favors with people on a one-to-one level, but he is helping others in the expectation that others in his peer group will reciprocate those favors. His wealth of *social capital* more than offsets his financial limitations.

Social capital can also be looked at from a group perspective. The social capital of a group refers to the collective value of all social networks and the inclinations that arise from these networks to do things for each other. Social capital is the application of reciprocity in social networks, where individuals are acquiring social capital by investing time, labor, knowledge, and other nonmaterial goods in their community in the expectation that the community will pay them back in kind.

Harvard sociologist Robert Putnam defines social capital as including aspects of social organizations, such as trust, norms, and networks that can improve the efficiency of the organization by facilitating coordination.[8] Another way to define social capital is to look at it as a stock of emotional attachment of individuals to a group. Social capital can be positive or negative. Horizontal networks of individual citizens and groups that enhance community productivity and cohesion are positive social capital assets, whereas self-serving exclusive gangs and hierarchical patronage systems that operate against communitarian interests can be thought of as negative social capital burdens on society.

Meritocracy, Consistency, and Transparency Build Social Capital

There are different ways to regenerate social capital that can be instituted at the organizational level but target the individual. In the context of virtual communities, the first step toward building social capital is to improve the education about, and awareness of, virtual teamworking in cyberteams. Knowledge of Internet etiquette and practice in Internet technologies are prerequisites for effective participation in virtual teams. Improved awareness, access, and education on virtual teamworking will lead to a more participatory workforce. A second step is to encourage increased participation in activities outside of the "core" business and to back community-oriented workplace practices. Deconstructing huge business units and creating small and flexible teams comprising a "we-feeling" supports this more active involvement of individuals into non-core activities. Supporting the creation of social capital also means that firms should act responsibly toward their employees'

■

Sidebar 4.3
A Broader View of Social Capital

The World Bank takes a very broad view on social capital, in which "social capital" refers to the institutions, relationships, and norms that shape the quality and quantity of social interactions.[1] In this view, social capital is not only the sum of the institutions that underpin society—it is the very glue that holds them together. Social capital is a company's "stock" of human connections, such as trust, personal networks, and a sense of community; it is so integral to business life that, without it, cooperative action and productive work are impossible.

Almost every managerial decision—from hiring, firing, and promotion, to implementing new technologies, to designing office space—is an opportunity for accumulating or losing social capital. To grow social capital in an organization means hiring and encouraging people who fit the values and culture of the organization, and creating an environment in which social capital will build. To do this, companies must take active steps to develop trust, networks, and communication. The benefits of this approach will be better knowledge sharing, lower transaction costs, lower turnover of key employees, better coherence of action due to organizational stability, and more shared understanding. There will also be more creativity if people are allowed to experience the intrinsic pleasures of taking the future into their own hands. Commitment and cooperation will increase, and customer needs will be more intelligently answered.

1. See http://lnweb18.worldbank.org/ESSD/Sdvext.nsf/09ByDocName
/SocialCapitalInitiativeWorkingPaperSeries.

■

families and community commitments and encourage other employers to follow their example. For instance, accommodating part-time work increases social capital, since part-time work increases one's exposure to wider social networks while leaving enough time to pursue other interests outside the workplace.

Wealth in social capital in an organization benefits both the individual and the group. This means that collaborative action and coordination is made much easier in an organization where high social capital has been accumulated. An organization whose members have accrued high levels of social capital provides a fertile nurturing ground for COINs, because these organizations exhibit strongly three characteristics linked to the accumulation of social capital:

- *Meritocracy*—organizations that reward and promote people solely based on merit
- *Consistency*—organizations that behave in a predictable way, governed by a mostly unwritten ethical code
- *Internal transparency*—sharing all the knowledge needed to make educated decisions among all members of a team

These three characteristics define an environment for today's organizations and companies to become a greenhouse for COINs. In combination, they form a powerful foundation for a collaborative culture where COINs will flourish.

The other aspect of COINs we should explore before looking at the specific cases in the next chapter are the "small worlds" in which COIN members communicate. Like the ethical code, the small-world networking structure makes it possible for each team member to be reached quickly and makes COINs flourish. This communication network structure is resilient to the loss of peripheral team members and scales well if new members join the team. Peripheral COIN team members are linked by a tightly connected core team, the "hubs of trust."

COINs Are Small-World Networks

I discovered how small the world in which we live really is on a cold winter evening in a delightful bed-and-breakfast inn in Cambridge, Massachusetts. I was chatting with another guest, who turned out to be a computer science professor from Helsinki, Finland. I was even more surprised to learn that we had a common friend in Helsinki. He was a colleague of mine working in our company's Finnish office, and he was also a former student of this Finnish professor.

The sociologist Stanley Milgram conducted a famous experiment in which he asked 50 people in the U.S. Midwest to send a letter to a final recipient whom they did not know in the U.S. Northeast. The Midwesterners were not allowed to mail the letter directly to the recipient, but had to forward it to another person whom they knew on a first-name basis, and whom they thought might be closer in some way to the final recipient of the letter. Each intermediary recipient was supposed to repeat this experiment until the letter would finally reach its destination.

To Milgram's surprise, it took only six steps, on average, for a letter to reach its destination. He concluded that the United States is indeed a "small world," with the population surprisingly well connected by a social network.

Duncan Watts, a mathematician and Columbia University sociology professor working with Stephen Strogartz, took up Milgram's ideas to define the mathematical properties of small-world networks (see figure 4.1).[9] The small-world network of me and the Finnish professor had a distance of only two, meaning that relaying Milgram's letter to our common friend in Helsinki would have been enough to get it from Cambridge to Helsinki.

At first glance, the "chance encounter" in the Cambridge inn came as a huge surprise, but it is less surprising than it seems. First, the chance of meeting another technically inclined person working with computers is quite large in Cambridge, which houses two of the world's leading research universities and is surrounded by dozens of other colleges and universities in and around Boston. These institutions have spun off hundreds of technology startup companies in the greater Boston area. Bumping into another technology person in Cambridge is not uncommon. Second, both the Finnish professor and I work as leaders in our respective communities and have large social networks in different geographic regions populated with people with similar profiles. A person with a university degree in computer science in Helsinki could well have taken classes with the Finnish professor and then gone on to work for an international consulting firm. The chances were much higher than average that, if I know an information technology guy in Helsinki, so does the professor.

COINs make direct use of the small-world property. With usually much less than six degrees of separation, COIN members can reach each other. In a well-functioning COIN, the average degree of separation is usually between one and two, meaning each team member can reach anybody else in the team directly or through a mutual friend.

COINs Are Scale-Free Networks

Notre Dame researcher Albert-Laszlo Barabasi's effort to map out the World Wide Web revealed that the Web follows what he terms a scale-free network distribution pattern characterized by both a small number of Web sites with many links, the "hubs," and a large number of sites with only a few links. This

■

Sidebar 4.4
Building Online Trust

Noted philosopher and ethicist Francis Fukuyama defines trust as "the expectation that arises within a community of regular, honest, and cooperative behavior, based on commonly shared norms, on the part of other members of that community."[1] The other term that Fukuyama uses in the same context is "spontaneous sociability," which is the ability to form new associations and to cooperate within the terms of reference they establish. If people who work together trust one another because they are all operating according to a common set of ethical norms, their spontaneous sociability will be much higher.

COIN members usually have high spontaneous sociability. The main glue that holds them together is the network of mutual trust. Trust can only be maintained if there is an agreed-upon code of ethics, such as the Tao of COINs.

A team of Finnish researchers looked at how teams of globally distributed software developers collaborated.[2] They identified a three-step process of building online trust:

I. Develop *familiarity* by having face-to-face meetings, workshops, and team-building exercises.

scale-free distribution pattern is combined with a small-world architecture that supports short routes through large-scale networks—which is why, for example, computer viruses and other disruptions spread so rapidly over the Internet. This phenomenon also accounts for the accelerated distribution of pirated content and networks' vulnerability to cascading failures. The advantage is that scale-free networks are very resilient against random failures. As long as the hubs are online, the main linking structure of the network remains intact, and all nodes can be reached by short network traversals.

COIN networks are also scale-free. There are a few extremely well-connected hubs that are crucial for successful operation of the COIN. As long as arbitrarily large numbers of peripheral people join and leave the COIN, the group continues to work just fine. If, on the other hand, just a few of the well-connected hubs in the core team leave, the COIN ceases to

2. Develop *confidence* by having co-located training and by making roles in the distributed teams explicit.
3. Maintain *trust* by keeping information flowing.

Meeting face-to-face is still the fastest way to build trust. In cases where it is not possible for global teams to get together physically, there are various substitutes. If all parties involved deliver obviously high-quality work, trust is built without meetings. But this process takes far more time than an initial face-to-face meeting, as team members have to let their work literally speak for itself. Additionally, if the team members come from different cultures, it can be hard to define a common language. In the software industry, for example, programmers from India, China, or the Philippines frequently work together with project leaders from the United States or Western Europe. Different attitudes toward issues such as praise and rewards, work schedule and quality, and loyalty within family can raise serious roadblocks in the effort to build distributed trust. Additional obstacles arise if both sides are not given enough information about the project, the tasks to be done, how the work and responsibilities are divided between sites, and what kind of quality is expected. Under those circumstances, lack of communication will lead to mistrust. In order to prevail over initial obstacles and to overcome prejudices, chat can be useful, as it allows asking questions and getting immediate feedback.

1. Fukuyama, *Trust: The Social Virtues and the Creation of Prosperity* (1995).
2. Paasivaara, Lassenius, and Pyysiäinen, *Communication Patterns and Practices* (2003).

operate normally, and may even dissolve. The higher the "small-world" and scale-free properties of a community are, the more robust it will be against disintegrating if some of the members leave. COINs exploit these small-world and scale-free properties to form a network of trust.

Initially, new members of a COIN will come with a preassigned level of trust, based on familiarity, reputation, and quality of available information and external recognition. Once new members have joined a COIN, they will develop and grow their level of trust in others and the trust that other COIN members place in them. Speed and level of growth is based on integrity and competence of the other COIN members they interact with, the quality of information access and communication flow, the intensity of the community building process, and the external perception und support of the COIN.

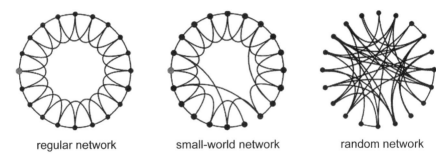

| regular network | small-world network | random network |

Figure 4.1. Regular, small-world, and random networks. Courtesy of Duncan Watts and *Nature*.

Hubs of Trust in Small-World, Scale-Free Networks

A network of trusted hubs operates far more efficiently. Thanks to a small-world and scale-free structure, it offers an extremely scalable and robust way of efficient virtual collaboration. Figure 4.2 illustrates the "small-world" property of the network. If John wants to reach Mary in conventional organization A, he has to pass by two intermediaries—that is, his path length to Mary is 3 and therefore his degree of separation from Mary is 3. With a small-world structure such as in organization B that allows for direct connections between any two members of the organization, the degree of separation is only 1.

Figure 4.3 illustrates how the small-world structure can be amended to become scale-free. Network C displays a scale-free network with John and Mary acting as hubs. Bill, Fred, and Sue can only access the network through John and Mary. This network is robust and works fine as long as only peripheral people such as Bill, Fred, and Sue are added or removed. The network breaks, though, if one of the hubs, such as John or Mary, leaves.

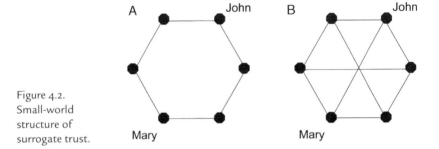

Figure 4.2.
Small-world
structure of
surrogate trust.

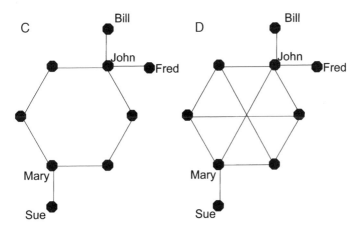

Figure 4.3.   Small-world, scale-free network of surrogate trust.

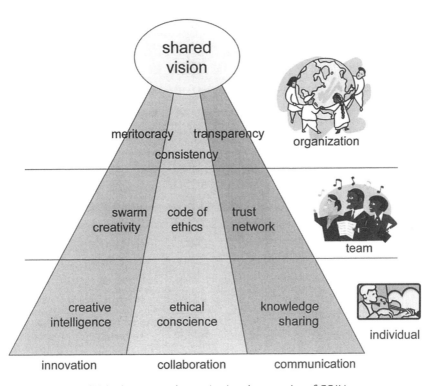

Figure 4.4.   Individual, team, and organizational properties of COINs.

The scale-free network C in figure 4.3 becomes even more robust and efficient if it is made a small world in D by adding direct connections between non-connected hubs. The connectivity between hubs John and Mary is now guaranteed, even if one of the other hubs in the circular path between John and Mary leaves. This means that a network of surrogate trust is ideally structured as a scale-free small-world network, with the hubs of trust acting as cornerstones of the network.

The Deloitte Consulting e.xpert network is an excellent example of such a hub and spoke scale-free small world where trust is given by proxy. It is described in detail in the next chapter, where you will see a graphic of the actual communication flow in the e.xpert community (see figure 5.2).

Figure 4.4 pulls together what we've learned about COINs in these chapters. The pyramid illustrates the core properties of COINs on the individual, the team, and organizational level. We've seen that meritocracy, consistency, and transparency comprise the defining elements of an organizational culture for COINs. We have identified three main organizational characteristics of COINs: swarm creativity, an ethical code, and a small-world network of trusted relationships among team participants. And we've observed, at the individual level, that individuals working in COINs possess particular cognitive and emotional characteristics that define the typical personal traits of collaborative innovators. In chapter 5, we will look at some real-life COINs in practice.

CHAPTER 5

Real-Life Examples

Lessons Learned from COINs

The lessons learned by participating in collaborative innovation networks (COINs) apply to every company and organization in daily business life. Companies such as DaimlerChrysler, IBM, Union Bank of Switzerland, Intel, and Microsoft, and organizations such as the United Nations, have harnessed the same principles with great success, leading to innovative new products and making existing processes more flexible, efficient, and agile.

The most effective lessons are those we learn directly. Accordingly, the cases presented here are all personal experiences I have had as a member of different COINs. The cases take us inside COINs to see how they work and what they can tell us about success for other enterprises. The descriptions in this chapter augment the initial presentations of the cases in chapter 2, where the concept of swarm creativity was introduced, with discussions of the ethics of COINs and the importance of "small-world" COINs . The cases also provide specific references for the "how-to" section that comprises the appendixes.

A Three-Point Process Built on Two Pillars

Each case here displays the typical pattern of COINs. A group of intrinsically motivated people got together to work toward a shared goal, starting on a shoestring budget and without organizational blessing. When management finally embraced the project (in those instances where management existed), the COIN was transformed into a project team or business unit. None of these COINs could have existed without the Internet, but what made the difference was not the technology but the *team*. With one exception, each of these COINs succeeded.

Each case also displays a similar pattern of progress. A three-step process of *discover-develop-disseminate* or *create-collaborate-communicate* (steps that correspond to COIN member roles we learned about earlier) unfolds on a foundation of two pillars: culture (with similar elements in every COIN) and heavy use of Internet technology. First, let's look at the process in brief.

- *Creation.* At each COIN's genesis, a small core team came together with ardor and enthusiasm, inspired to make a shared vision come true. The makeup of the team, or the specific type of organization or business sector, didn't matter. Rather, what mattered was the common shared vision, which drove the team members to work together to create something original.

- *Collaboration.* After the initial step of creation, the creative idea was taken up by a larger team that worked together in truly collaborative fashion. Each COIN began to add peripheral members. This was the critical stage for acceptance of the idea within the organization hosting the COIN, so the meritocratic mindset proved crucial. This did not mean that hierarchy was absent from the COINs; on the contrary, people selected their own leaders. But everyone could speak, everyone was treated with respect, and even the most junior team member had the right and responsibility to come up with constructive criticisms or new ideas.

- *Communication.* Once each core team achieved its first goals, successes had to be communicated to a wider audience.

The process was built on two foundational pillars.

1. *Culture.* COINs succeed because they are able to foster the right culture of transparency, consistency, and meritocracy. This was clearly the lesson in the cases you'll read about below.

2. *Technology.* The Internet was the main technological enabler of these COINs, providing instantaneous and asynchronous reach on a global scale. The technology also stored a permanent trail of what had been happening in the COIN. Absent the communicative capabilities the Internet affords, the successful outcomes of several of the COINs you'll read about would have been impossible.

Figure 5.1 visualizes these five components—three process steps and two pillars—of successful COINs.

Let's look first at the DaimlerChrysler COIN.

DaimlerChrysler's e-extended Enterprise (e3)

I introduced COINs in chapter 1 with the phone call that led to my involvement with the global procurement group of the then newly merged DaimlerChrysler. Two key lessons emerged from the DaimlerChrysler e[3] project: (a) what makes COINs so effective is they *communicate efficiently by sharing knowledge within the team,* and (b) COINs *build and maintain distributed trust.*

For businesses, these lessons lead to a direct benefit: the knowledge sharing inherent to COINs lowers transaction costs.

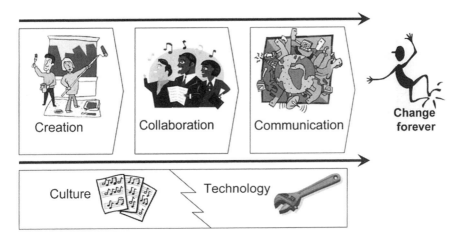

Figure 5.1. Components of the COIN-driven innovation process.

Project Genesis and Overview

At the end of 1999, DaimlerChrysler decided to develop an e-business strategy that would streamline its Global Procurement and Supply operations, optimize transaction processes, tighten control of materials flow and inventory, and improve the transfer of information throughout its supply chain. The solution needed to be one that enhanced relationships with suppliers, increased efficiencies, reduced new product development time, and cut expenses. To achieve these goals, the company launched the e-extended Enterprise, or "e³," initiative, which today links DaimlerChrysler with its supplier partners in a supply community.

DaimlerChrysler's competitors were also busy leveraging e-business technologies in automotive procurement and supply chain processes. So the company wanted a solution that was *unique*.

As discussed in chapter 1, the e³ project began in January 2000. Approximately 15 people from the Global Procurement and Supply function worked full-time, supported by members of the information technology (IT) organization; employees from the business units worked part-time, as required. A team of 15 external consultants from the United States and Europe reinforced the project team. This entire group operated as a global COIN.

In the e³ initiative, DaimlerChrysler examined and reevaluated its organizational structure as well as the roles and responsibilities required for new and redesigned Web-based business processes. The company also created new processes around Web publishing, transactions, and integration of sales and marketing, and developed new technology components—from the user interface to back-end integration. The new Global Procurement and Supply solution that was developed integrates legacy systems from both Daimler and Chrysler, and it addresses the various technologies of its supplier partners; this system has been rolled out worldwide. And while it was not planned at the beginning, the COIN came up with a highly innovative idea: to join forces with DaimlerChrysler's competitors by creating the Covisint online marketplace, a worldwide automotive exchange. The Covisint exchange—which includes Ford, General Motors, Renault, and Nissan as major partners—is now an integral part of DaimlerChrysler's procurement solution.

Chapter 2 explored the swarm creativity that was so vital to the success of the e³ project. Let's look at two other key aspects of COINs that were discussed in chapter 4.

The e³ Ethical Code

The concept of an ethical code, so important to a COIN's success, was a central element of the DaimlerChrysler project. People working on the e³ project became committed team members by experiencing the COIN attachment mechanisms, namely, investment, communion, and transcendence. In workshops held in Michigan and in Stuttgart, Germany, the team developed a strong "we-feeling" of belonging to a select group that was on a mission—to change the procurement processes of DaimlerChrysler. Working long hours and giving up functions outside the e³ project itself, members grew a strong attachment to the community. Important decisions were made in "transcendent behavior" not by the team leaders, but in full-team workshops.

Communication among team members was very direct, and social capital in the team was high. People treated others as they themselves wished to be treated. In workshop presentations, immediate and honest feedback was given to presenters. Although the team numbered up to 30 people, all were directly connected, communicating not through project and subteam leaders but by making information accessible for whoever needed it to get the job done. The degree of sharing became very high. Once trust was established in face-to-face meetings, open knowledge sharing took place over the Internet. A Lotus Notes-based Web site served as the main knowledge repository where relevant information, even highly sensitive documents, was accessible to all team members independent of whether they were DaimlerChrysler senior or junior staff or external consultants. All project members received a daily project update. The degree of individual interactivity was also very high; team members interacted intensively, and requests—whether made face-to-face, by phone, or by e-mail—were usually answered the same day.

The culture in the mixed team of German and U.S. DaimlerChrysler employees and external consultants was governed by the Tao of COINs. While team leaders and senior experts were highly regarded, all team members were treated respectfully. We employed a fact-based, non-hierarchical communication style. Senior experts took the time to introduce and tutor novices, who then quickly became valuable team members. The project and team leaders came from a wide range of different hierarchy levels, nationalities, and ethnicities, demonstrating the meritocracy of the project. We had defined clear rules of team-based and hierarchical decision-making, thus demonstrating consistency; we also shared all knowledge on our project Web server and by mailing lists, therefore demonstrating transparency. Working together in swarm creativity, with our mutual behavior defined by the Tao of COINs, members also connected with each other in an efficient communication network.

The e³ Project's Small World

The DaimlerChrysler e³ project also exhibited a typical scale-free, small-world structure. The overall team comprised, at times, up to 30 people from DaimlerChrysler and different consulting firms. A larger group of more than 60 people from different parts of the DaimlerChrysler organization complemented the team, participating temporarily at different phases of the project as outside subject matter experts. The team leaders from DaimlerChrysler and the consulting company, together with the different subteam leaders, acted as hubs of trust in the communication networks of the larger project team. The members of this core team got to know each other well at early face-to-face workshops in Michigan and Stuttgart; once trust was established, team meetings were teleconferences. And because the e³ project leaders acted as hubs of trust, senior DaimlerChrysler managers on both sides of the Atlantic who were not part of the e³ project team agreed to share sensitive data. By gaining the confidence of senior procurement managers in Stuttgart, the e³ team leaders became bridges of trust and could convince those managers to share data that they had originally wanted to keep secret with their counterparts in Michigan. Moreover, junior e³ team members who had no opportunity to cross the ocean to get to know their counterparts trusted each other remotely and shared sensitive data because their local team leaders were acting as hubs of trust on their behalf.

Lessons Learned

My own involvement in the e³ initiative was a great learning experience in the workings of a COIN operating on a global scale in a major multinational corporation. Overcoming initial distrust between team members from both sides of the Atlantic and successfully establishing teamwide trust fostered a spirit of true collaboration within the project, which resulted in a huge efficiency improvement. Three areas in particular were essential for effective collaboration: (a) efficient sharing of knowledge by communicating the right message tailored to the audience; (b) flexibility to change one's role in the course of the project; and (b) building and maintaining trust virtually, without meeting face-to-face.

People and company culture are the critical factors in large-scale projects such as the e^3 initiative. A principal challenge was to enable DaimlerChrysler to "think the unthinkable." At the outset, people were apprehensive, and apprehension hampers innovation. Staff members were worried that projected savings and streamlining of procurement processes might lead to elimination of their jobs. So it was crucial to communicate openly what changes would occur and what opportunities would be afforded to individuals.

A major advantage of any project based on Internet technology is that initial results are seen quickly, so people can relate to, and identify, changes immediately. The e^3 project made full use of Web technologies, publishing position documents and using Lotus Notes-based collaboration tools. The guiding principle was that there can never be too much communication, and project management sent out daily project updates to all team members. At first, this led to information overload. Over time, a more personalized e-mail service with abstracts was provided, with the full contents stored on the project intranet Web site to reduce the cognitive overload.

Despite apprehensions, some communication glitches, and all the other obstacles, we learned an important lesson: *COINs communicate efficiently by sharing knowledge within the team.* We also learned that COINs operate as what are called "X-teams" (see sidebar 5.1).

Another crucial aspect of the e^3 project was to bridge cultural differences between the procurement and supply departments in the United States and Germany. Initially, the cultural differences were profound. For instance, in the United States, mistakes are accepted as a natural part of the process, and speed is seen as an essential driver. In Germany (and Europe in general), the tendency is to accelerate efforts only when some positive results have been achieved; this appears to make the process slower. To establish a common pace, people from the United States were brought to Germany and vice versa. They reached a common understanding, and both sides agreed to try to let go of these expectations, which amounted to cultural barriers.

Initially, DaimlerChrysler executives from both sides of the Atlantic were quite reluctant to share confidential data. Once trust was established in face-to-face workshops, collaboration over long distance became much

Sidebar 5.1
COINs as X-Teams

The DaimlerChrysler e³ initiative was at the forefront of strategy and technology, and one of its features was that team composition changed considerably over the project's lifetime. Working together was a continuously changing group of experts from the company and external consultants. MIT professor Deborah Ancona calls this an "X-team"—a team with permeable boundaries and changing roles, in which members move in and out, and in which the focus is on getting the job done.[1]

The idea of changing roles is clear from DaimlerChrysler's story. Team members had a high level of motivation and excitement; they felt they were pushing limits and could have a real impact on DaimlerChrysler's future. But roles weren't always clear, and team members sometimes misunderstood assignments or duplicated tasks. Team members also took on different roles with different levels of engagement over the course of the project. For example, during the first nine months, project and sub-project leaders changed frequently because of the need for different areas of expertise.

DaimlerChrysler team members from Stuttgart and Auburn Hills joined and left the team frequently or changed their level of commitment. Project management began to see the need to make roles and responsibilities more clear and accordingly assigned guides to introduce new team members quickly. The environment changed daily, so it was key to keep team members informed in this volatile atmosphere by immediately communicating personal and project goal changes to the entire team in a condensed and personalized way.

Ancona distinguishes as well between three levels of commitment, which we saw in the DaimlerChrysler X-team and in the Swiss private bank (discussed in this chapter) X-team: core, team, and extended team. Anecdotally, I've heard these called "pigs" (who provide the "bacon" or the "gameskin"), "cows" (who give the best they have, which is milk), and "chickens" (who provide the occasional egg).

The e³ project afforded valuable insights into building a distributed team operating as a COIN. Once operations as an X-team were understood and set up and distributed trust was established, the team could operated much more efficiently by maintaining high transparency through continuous knowledge sharing and communication.

1. Ancona, Bresman, and Kaeufer, "The Comparative Advantage of X-Teams" (2003).

more productive. For instance, video- and Web-conference meetings and workshops could be conducted productively. After overcoming the initial resistance, and once relationships were established, information was shared openly at the virtual meetings. It took the establishment of mutual trust before team members would offer their insights and reveal confidential information to their counterparts on both sides of the Atlantic. The clear lesson is that *COINs build and maintain distributed trust.*

Of course, the other lesson of the e³ project is one of the benefits from chapter 1: *COINs make organizations more collaborative.* The initiative led to a separate company, owned jointly by DaimlerChrysler, General Motors, and Ford.

Deloitte Consulting's e.Xperts

Trust increases knowledge sharing. Because consistent behavior—according to a mutually agreed-upon code of ethics—increases trust, COIN members collaborate more efficiently and have less need for time-consuming coordination meetings, therefore working more productively. A COIN-driven activity such as the e.Xpert virtual e-business consulting practice of Deloitte Consulting in Europe demonstrates how networks of trust boost product quality.

Background and Project Genesis

I joined Deloitte Consulting in 1999 to help build up the firm's European e-business practice. In fact, I was with Deloitte when the phone rang inviting me to get involved in what became DaimlerChrysler's e³ project. At the time, Deloitte Consulting was a distinct part of Deloitte Touche Tohmatsu (DTT), a global firm of 120,000 practitioners delivering assurance and advisory, tax, and consulting services to clients in 140 countries. DTT established the consulting arm as a distinct organization in 1996, with major offices in the Americas, Europe, Asia/Pacific, and Africa.[1]

Deloitte Consulting grew, thanks in large part to the firm's approach—a more collegial and flexible consulting style than found in other large consulting firms—and to integration across the consulting practice. Major contributors to revenue growth included e-business; the development of key services, including customer relationship management, enterprise resource planning, supply chain management, and outsourcing; and a global expansion strategy. Focusing on seven business sectors—manufacturing, public sector, consumer business, health care, financial services, communications, and energy—the firm pulled together teams across the four service lines (strategy, process, technology, people) for each engagement, with the aim of helping organizations achieve strategy, process, technology, and people transformations. In 1999, Deloitte Consulting expanded its portfolio of service offerings to include the latest developments in e-business technology, such as e-procurement, portals, and other collaborative tools; e-supply chain optimization systems; and an integrated value chain approach.

When I joined Deloitte Consulting, my colleagues in the United States were already in the process of creating a dedicated e-business consulting unit they called dc.com, comprised of 200 consultants and partners working exclusively on e-business engagements out of offices in the Silicon Valley, East Coast, and selected other U.S. locations. We decided on a different approach in Europe—we would build up a virtual practice, the e.Xperts. We assembled a team of Deloitte Consulting practitioners working from offices throughout the continent.[2] We selected an "e-Champion" in each country to be the point of contact and leader of the national e.Xperts group. Our e-business leadership team embarked on a whirlwind tour to visit the national offices, educating practitioners about the e.Xpert concept. In most countries, we found a senior manager who volunteered to become the national e-champion and who served as a natural point of contact.

It was crucial that e.Xperts also remain full members of their national practices and of their competency areas, which included strategy, process, technology, and a category Deloitte called "people" (encompassing human capital, change leadership, and learning). Similarly, the e.Xpert process was self-selecting: motivated practitioners volunteered to become e.Xperts. We offered a structured training program, turning experienced consultants into full-fledged e.Xperts.

COINs Emerge

We wanted to build up a virtual e-business consulting practice across Europe, providing skilled resources to staff e-business projects, as needed. We also wanted to grow a core of Internet and e-business technology and strategy experts. These experts would be able to team with other practitioners, teaching them to improve client processes with their e-business skills. The goal was to recruit internally a core of 150 to 200 e-business specialists, with 80 percent of the consultants focusing on technology and delivery and the other 20 percent focusing on strategy.

Within a short period of time, a vibrant community of Deloitte Consulting e.Xperts was born, with whom we pursued multiple goals. We used the e.Xpert community as a recruiting tool to attract highly sought-after e-business experts, as well as a retention device for highly marketable staff with Web-relevant skills; we also made it very clear the e.Xpert organization was never seen as a new service line, but rather as a virtual organization across all existing service lines. The result was the development of a multidimensional matrix organization. The e.Xperts were a network of smaller national teams with peo-ple participating concurrently in cross-national subject matter expert practices (e.g., one group of people worked on Internet banking projects, while another worked on e-learning projects). All of these were COINs led by an e-Champion; the e-Champions themselves were forming a particularly close-knit COIN.

We fostered a sense of community by inviting all the e.Xperts to bian-nual meetings in central locations, such as Paris or Brussels, but all other collaboration was virtual. Leaders of the different COINs held regular Web meetings.

Hubs of Trust

One of the powerful discoveries made in building up Deloitte Consulting's e.Xpert community was about how people delegate trust. In cases where they had to work together with someone they did not yet know, people would rely on the judgment of people they trusted. And this reliance on other people's friends worked not only locally, but also extended into a pan-European network of trust by proxy.

Within the e.Xpert network, e-champions became the hubs of trust. The e-champions led various subcommunities (such as the national e.Xpert group in each country); there was also an e-champion for each industry and practice area (e.g., e-business consulting for financial institutions or the Java devel-opers practice group). The 24 e-Champions knew and trusted each other well, having become acquainted through regular pan-European e-Champion meetings. As senior members of the consulting organization, many had also worked together previously in consulting assignments.

The e-Champions formed a pan-European network of trust, which they were able to extend to other members of their organizations. This meant, for example, that an e-Champion in Lisbon might act as a reference for one of his team's e.Xperts, delivering a fair assessment of his or her skills to a project leader in Norway, who would send an inquiry through the national e-Champion in Oslo. The e-Champion network of trust was similarly effective in recruiting new e.Xperts, staffing projects with e.Xperts, and identifying the right people who might have personal relations to a potential client or business partner. The project leader in Norway was placing trust in the e.Xpert in Lisbon through two intermediate hubs of trust. The leader was using his personal network of trust, extending it to the e-Champion in Oslo, whom he knew, along to the e-Champion on Lisbon, whom the e-Champion in Oslo trusted, and finally to the e.Xpert consultant in Lisbon. This e-Champion network of trust allowed for extremely efficient operation at very low cost. Without the e-Champions, the project leader in Norway would have had to place his request with the European central staffing manager in London, who would then have used his skill database and contacts with conventional management to identify suitable candidates. The project leader would then have done multiple interviews to find the best candidate, leading to substantial communication and travel overhead. But the network of trusted hubs operates far more efficiently, offering an extremely scalable and robust way of efficient virtual collaboration.

Figure 5.2, produced by automatically analyzing the e-mail archives of the community using the temporal communication flow analysis (TeCFlow) system (detailed in appendix B), shows the actual communication flow in Deloitte's e.Xpert community. The people represented by the core team in the center are the most connected e-Champions, acting as hubs of trust. The coordinator of the community is in the center of the network and is communicating with everyone. The e-Champions are actively communicating among themselves. The other e.Xperts talk mostly to e-Champions and the coordinator, but have very little communication among themselves. This is a typical scale-free structure, with a highly connected center and a large number of peripheral people who are mostly connected to the center. As long as the e-Champions remain active, large numbers of e.Xperts can be

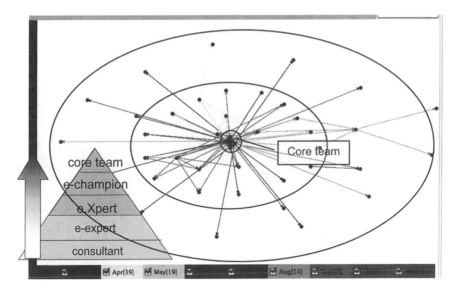

Figure 5.2. Small-world, scale-free communication flow in an e.Xpert community. The picture was produced with TeCFlow (see appendix B).

added or removed without any risk to the overall network. If just a few of the e-Champions leave, though, the e.Xpert network would dissolve. The central coordinator has a critical role; if he leaves, many e.Xperts would no longer be reachable. The e.Xpert network is also a small world: because of the high connectivity of the e-Champions, most e.Xperts have a degree of separation of two to three, which means that the path from one peripheral e.Xpert to another e.Xpert takes at most three steps: from the e.Xpert to an e-Champion, then to another e-Champion, and then to the destination e.Xpert.

Lessons Learned

That *COINs communicate efficiently by hubs of trust* was a key lesson learned in the Deloitte Consulting example. The fully committed approach to virtualization, coupled with a sharp focus on building up high levels of trust within the COIN, was a critical success factor. The main elements of building that trust were a high degree of transparency with respect to recruitment and reward of the e.Xperts, as well as a tightly woven network of e-Champions. The e-Champions knew and trusted each other personally and were able to convey their level of trust to e-Champions from other countries and other service lines, as well as to the members of their own local e.xpert teams. (This aspect of the e.xpert network is discussed as an application of the TeCFlow tool in appendix B.)

The Deloitte story also teaches us that *COINs are highly agile and productive at a very low cost.* The flexibility and agility of the e.xpert concept allowed rapid reaction to external market changes. In the hectic e-business days of 1998 to 2000, the e.xperts were instrumental in successfully acquiring and delivering projects well in excess of $100 million. When the e-business wave cooled down in 2001, it was simple and straightforward to fold the virtual e.xpert community back into mainstream Deloitte Consulting. Some of the COINs making up the e.xpert community maintained a life of their own; they are still vibrant communities and in some instances have become service areas in their own right, such as the groups focusing on knowledge management, on e-learning, and on customer relationship management (CRM). The e.xperts were an extremely fast and flexible—as well as very cost-efficient—way not only to react to the torrent of e-business, but also to be well ahead of the market. Deloitte Consulting Europe was able to nurture internal talent instead of resorting to expensive external hiring. This increased profitability and enhanced the firm's reputation in the market. As an added benefit, Deloitte's e.Xpert structure earned high ratings by industry analysts such as Gartner Group, Forrester Research, and IDC, making Deloitte Consulting one of the top-ranked e-business consulting organizations.

A project I was involved with as part of the Deloitte e.Xpert consulting practice provides another outstanding example of a COIN. While this next case involves the financial services industry, the basic principles of operating as a COIN are the same as in the automobile industry, for example our DaimlerChrysler case.

e-Banking for a Private Swiss Bank

In early 2000, e-business excitement had reached a climax. "Brick and mortar" companies were out of fashion and every conventional company had to have at least one e-business operation. In financial services, multifunctional "lifestyle" online banking portals were considered the holy grail of e-banking, where wealthy clients could obtain all services having to do with money in one virtual place. Deloitte Consulting's client in Switzerland, a private bank, had already developed its business case for such an online banking portal.

Specifically, the bank wanted to create a Web-based e-banking portal with leading-edge functionality for its private clients. The portal was to serve all three segments of the bank's clients: local retail clients, by far the largest client group in absolute numbers (but who contributed a small part to the bank's profits); the wealthy and affluent client group, which invested between $100,000 and $1 million and was the second-largest group in numbers (and highly profitable for the bank); and the high-net-worth individuals who invested more than $1 million, up to hundreds of millions of dollars.

The project was a complex mix of technology and business process reengineering. It afforded excellent opportunities to further hone the skills needed to get COINs in a commercial environment up and running. Unlike the DaimlerChrysler example, the challenge here was to establish and maintain local trust in a virtual team. It was crucial to create a common vision and goals among all team members, a goal achieved only relatively late in the project.

The e-banking project ultimately taught us many lessons. Some of the main lessons had little to do with the virtual nature of the project and a lot to do with commonsense project management. The bank project managers, as well as the external consultants developing the e-banking system, had accumulated dozens of years of software-development project management experience and were quite skilled in building similar systems. Nevertheless, we all fell into the usual traps of planning for optimistic assumptions and not building in enough contingency time for unforeseen circumstances. For example, due to erroneous manipulations, the project's whole hardware and software infrastructure, already completely installed, broke down in final testing and crashed not only the hard drives but also every backup. This meant that the full development, testing, and run-time environments had to be recreated and re-installed from scratch—resulting in a four-week project delay and adding another four weeks of development time for the system administrator and application development teams.

The major lesson was an old insight: that the commitment of team members is the centerpiece of project success. What was new here was the distributed nature of the project. While the bulk of the development was done on-site in the bank's offices, some of the software coding and customization of off-the-shelf software components was done in Italy, Belgium, Germany, Canada, the United Kingdom, and the United States. The biggest part of the team, which numbered between 30 and 40 people, came from the bank and from Deloitte; while English was the main project language, the native languages of team members included Italian, German, Spanish, English, and Flemish. Additionally, the cultures of the bank, the consulting company, and the freelance subcontractors were quite different. Team members varied widely in their level of project experience and skill sets.

Project leaders were aware of the importance of a unified team and organized monthly team-building events. While the core team of bank employees and consultants worked together very well from the beginning, full collaboration between the other bank employees and consultants only grew toward the end of the project. For example, handoffs of completed milestones and deliverables from the system developers to the bank's maintenance staff initially presented difficulties, as the bank employees shunned taking over responsibility.

A key lesson learned was *COINs can succeed only if the culture is right.* Over 14 months, we observed three levels of team integration that taught this lesson.

First was *building the team,* where the rules of the game were defined and the project culture created. We learned a collaborative working culture must be established in the beginning. This clarifies the work ethic (e.g., hours worked per day, rules for teleworking, etc.). Rules for communication among team members and with the outside world need to be established, and the role of each team need to be defined.

Second was *working together.* Here we saw the project team as an X-team (see sidebar 5.1 description). Because the project was using leading-edge technology, a continuously changing group of technicians was working together on making the different software modules interoperable. Team members took on different roles with different levels of engagement over the project's lifetime. For example, one of the initial lead software architects subsequently became developer of a software module, then head of testing, then main architect of the next release, at ever-increasing percentages of his overall capacity. Rather than trying to stick to a rigid team structure with well-defined roles, we continuously planned for these changing roles in an X-team—and this allowed us to make much better use of resources and saved project costs. (Sometimes, however, even a "dream team" can fail, as sidebar 5.2 explains.)

Third was *to optimize the knowledge flow.* Project members began to operate as a unified team only toward the end of the first project phase. At that stage

Sidebar 5.2
Learning about COINs from a Failure

This is the story of a failure. *How* we failed provides some valuable insights into what really matters in a COIN.

Cybermap, a concept I developed while a postdoctoral student at the Massachusetts Institute of Technology, was a system to create automatically "geographic maps" of the World Wide Web, or of any other large collection of documents. For a while, I pursued Cybermap as a purely academic interest, working in my spare time with a computer scientist at Dartmouth College to refine the concept. He and I were forming a small COIN core team, collaborating virtually between New Hampshire and Switzerland most of the time.

In 1997, we decided that the time had come to start a company and turn Cybermap into a full commercial product. We convinced four colleagues to join us and reinforce our technical and financial team. All of us had worked together successfully in the past.

It was a "dream team": we already knew and trusted each other, we had excellent and complementary skills, and we had high confidence in each other's capabilities. Our product was truly innovative and without close competition. And yet we failed. Why? We badly messed up the transition from informal team to commercial company.

The transition problem exhibited itself primarily in two areas: loss of focus and loss of interpersonal trust. Losing our shared vision was a main roadblock. Some of us pursued other activities in parallel to building up our company. The company was my brainchild, but I did not devote enough attention to it at a critical time in its growth. Over time, we lost our shared vision. Each team member developed his own sense of what we were trying to achieve, and these goals moved further and further apart.

We also failed in another important way: One important aspect of the transformation from informal team to commercial company is that team members begin to be paid not only in peer recognition, and by having fun, but also in real money. The failure to give sufficient attention to *how* team members would be compensated helped reduce the mutual trust we shared, and loss of trust destroys a COIN.

While we had started out as a COIN, we never were able to overcome the initial transition phase of getting the core team fully operational and turning the loose research team into a fully operational COIN. It seemed as if we had done everything right in the beginning and everything wrong in the end—squandering the fortunes of our company by losing our shared vision and mutual trust.

in project life, the number of people working on the project had dropped from more than 40 to just nine. Team collaboration had been optimized, with those nine people working together far more productively than had the much larger group. In intensive knowledge management workshops, we had repeatedly questioned everyone's role well as the design of the system, resulting in a redesigned knowledge flow leading to more efficient collaboration. We had one common, overarching goal: getting the system up and running as quickly as possible. We also now knew the strengths and weaknesses of all team members, and nonproductive members had been weeded out. Virtual trust had been established for productive virtual collaboration.

Linked to this third step was another key lesson—that *optimizing knowledge flow comes before technology*. In the e-banking project, we reconfirmed the main insight of what is called "agile software development"—that it is the human being who is the key to success, and not processes, tools, documentation, contracts, and plans (see sidebar 5.3).[3] This does not mean, of course, that processes, tools, documentation, contracts, and plans are unimportant, but rather that collaboration within the project team must come first. This is the key point for the success of any COIN.

Another bank project offers an illustration of swarm creativity—developing the Interoperability Service Interface (ISI) for Union Bank of Switzerland. It came after another COIN experience at the same bank.

Union Bank of Switzerland (UBS)

My first contact with the power of COINs at UBS came with the development of the bank's intranet. I was a software manager and had started my own "skunk works" to create the "Bank Wide Web" for information sharing within the bank. At first, my boss frowned upon the idea of using an "academic" system—in this case the World Wide Web, which was not nearly as ubiquitous as it is today—for productive, business-critical applications; he instead wanted me to use a commercially available document management system. But I hired a summer intern from the Massachusetts Institute of Technology, and this intern built a prototype and made an evaluation comparing the two systems that clearly pointed out the advantages of the Web.

Still, my boss was reluctant. What finally turned him around was a symposium he attended of IT industry analysts from Gartner Group. Gartner

Sidebar 5.3
What We Learn from Agile Software Development

Agile software development is based on four principles that correspond well to what we've learned about COINs:

1. *Individuals and interactions over processes and tools.* While there have been numerous proposals for the ideal software development process and even more proposals for tools to build software, none has proven to be the "silver bullet." While it makes sense for a team to define a development process and select a set of tools, the people who apply those tools make all the difference. An experienced software developer who knows what she or he is doing and has accumulated years of experience is worth much more than the best tools and processes in the hands of a novice.

2. *Working software over comprehensive documentation.* While it is important to document what has been developed to capture explicit and tacit knowledge, it is the running system that delivers the best description of what has been achieved. It is key to document the inner workings of the system, knowing that the documentation and the running system almost never will be in perfect synch. The ultimate documentation is a flawlessly operating system. In our e-banking project, we produced elaborate documentation, which was always one generation behind the actual system.

3. *Customer collaboration over contract negotiation.* While it is important to negotiate a contract with customers to manage their expectations, collaboration with, and not against, clients is key. If the customer is involved in the development of the system and feels like part of the team, it will be much easier to bring up and resolve problems as they arise. Open communication and a trust-based relationship with the customer is the best insurance against unpleasant surprises. In our e-banking project, the relationship between Deloitte and the bank was defined in a complex contract, but we were able to complete the system only when we replaced strict adherence to the contract with close customer collaboration.

4. *Responding to change over following a plan.* Starting a project with a clear plan that outlines major milestones is important, but even more important is to know that there will be changes to the plan over the course of the project. This means that the project plan needs to be adjusted periodically in response to change. In our project, the deliverables defined in the contract started to change before the contract was even signed: we were adjusting the system specification in biweekly change board meetings with client and consulting team leaders.

strongly recommended using the Web for information sharing. My case was supported by the fact that, by then, the U.S. arm of our company had already started to deploy the Web internally. Suddenly, I was getting complaints for being too hesitant in rolling out the Web! After about six months, the Bank Wide Web was officially approved and converted first into a project team, later into an official business unit.

As is typical with COINs, the life cycle began with an informal group of people sharing the same goal and vision. The hosting organization only acknowledged the value of its COIN after external recognition. Once the organization hosting the COIN recognized its value, the COIN was converted into a sanctioned organizational form.

In the case of the Bank Wide Web, as the members of the team were working, we would go out to lunch together or engage in other joint social activities. When we were having lunch or drinks in a restaurant, we would talk frequently about the latest technical developments of the Web and of Java, furthering our knowledge and learning from each other. This built up a high level of mutual trust, enough so that we could discuss more personal matters such as external job opportunities of interest to a team member. Because of our excitement, we acted as ambassadors, seeking to persuade the rest of the bank to start using the Web. Because we were convinced of our cause, we acted as efficient communicators. Rather than trying to reach as many people as possible, we tried to reach the right people, knowing that quality of interactions is more important than quantity of contacts made—after all, networking is not an end to a means, but a means to an end. Just getting to know other people for the sake of knowing more people is not much help in disseminating new ideas, and might even be counterproductive. Not every innovator is a natural salesperson who can easily approach strangers, turning them into close friends in the course of a conversation. And not all members of a COIN are born communicators, but in an effective COIN, all members become missionaries. The rapid distribution of the new idea is a consequence of the power of the communal vision that cyberteam members radiate and transmit to their environment. If people are fully convinced of the merit of the new idea, they act as communicators and ambassadors for their COIN.

An even deeper lesson into the workings of COINs at UBS came later, with the development of the ISI. The bank's goal was to change its IT system to be able to store all banking transactions in non-condensed, so-called "atomic" format over a period of five years. Previously, data had been compressed to show only the customer's last few transactions and the current balance. On a more technical level, in the current IT system, data was accessed directly on the record level, meaning that the specific format and location of the data needed to be known to the application programmer. This not only required that data access routines be rewritten for each new program, but far worse—if the data format changed, all the computer programs accessing the data had to be changed, too.

When I joined UBS, my predecessor had already begun to look at a promising new technology called Common Object Request Broker Architecture (CORBA). But this technology was then still considered to be too immature and risky for use in a high-security environment such as the transaction backbone of a large bank. When I took over the project, my predecessor as project leader convinced me of the advantages of the CORBA concept over more conventional approaches that UBS was pursuing at that time. So we embarked on an internal marketing campaign, trying to convince senior management of CORBA's advantages. In various senior management meetings, I introduced the concept to my direct boss, his boss, and his peers. I also asked a colleague to build a proof-of-concept prototype. Within a few weeks, he had created a prototype compiler for the system we were proposing. This was something quite unconventional for a bank, as banks usually build their own transaction processing systems, but do not want to develop in-house tools for building transaction-processing systems. It was as if we were building a house: not just digging a hole and putting up concrete and brick walls, as usual, but first building the excavator to excavate the construction site and then the concrete mixer to mix our concrete. As there was no "excavator" or "concrete mixer" available for CORBA at UBS, we had no choice but to build our own. And as the first prototype worked quite well, I was allowed to hire a small team of contractors, soon to be reinforced with internal IT staff, to build a production system of our excavator and concrete mixer for CORBA.

Over the course of two years, we successfully introduced ISI as a home-grown version of CORBA at UBS. ISI has been well accepted at the bank, and it has even been marketed outside the bank to other financial institutions as a commercial product.

Through the ISI project, we learned the valuable lesson that *COINs drive collaborative creativity.* The main reason for the success of ISI was the availability of a large pool of potential members of COINs all sharing the creator's genes at UBS. There was a culture of openness toward new ideas.

This culture was not just internally known; the bank had built an external reputation as a successful adopter of leading-edge technologies. The bank's IT department at that time even had its own applied IT research lab, where a University of Zurich professor on leave, together with a dozen researchers, was studying the applicability of groundbreaking computer science research results to the banking environment. Later, this professor actually moved into mainstream software application development at the bank, becoming a senior vice president and the head of the overall software application development department. This meant that we were able to build on a large pool of talented, motivated, and creative software engineers to recruit members for our "grassroots" project. The bank's reputation as a successful adaptor of "cool" new technologies had brought together enough people sharing the "creator's genes"—willing to take risks and invest working time and personal reputation into a new task outside of conventional territory.

We were forming a COIN, a team of people totally committed to the ISI concept. Our goal was to make ISI the main "glue" that would link together the distributed applications of UBS. We were also lucky in that timing and technology were right: The CORBA technology at the core of ISI was about to leave the "hype" stadium and to become really useful in a high-security production environment. There were now enough parts vendors of "nuts and bolts" for the bank to comfortably build its own concrete mixer and excavator.

Leveraging the Lessons Learned

COINs will happen, whether an organization likes it or not. Organizations have no choice but to accept that COINs are active within their organizational boundaries. It is therefore much better for the organization to be aware of this process and to leverage the creative output of the COINs to its advantage. By investing in COINs, organizations have a fast, flexible, and cost-effective way to innovate and pull ahead of the competition.

This book concludes with three appendixes that explain how to leverage the principles of COINs in organizations. The appendixes are a "how-to" guide, with tools. Each of the tools puts the principles we've learned to work and employs the most up-to-date advances in communications technology. Understanding some of that technology will help put the principles of COINs into a broader perspective—this is the subject of the next chapter.

Sidebar 5.4
A COIN That Truly Thinks Globally and Acts Locally

In 1992, the United Nations Conference on Trade and Development initiated the Trade Point Program (Trade Points) to "assist Small and Medium-Sized Enterprises . . . in overcoming informational, financial and logistic obstacles to increased participation in international trade, with particular emphasis on firms in developing countries." At first, Trade Points attempted to bring together under one roof all the services needed by exporters. Today, Trade Points enable small business owners in poor countries to use the Internet and Web to access computerized information on markets, potential clients and investors, and tariffs and trade rules worldwide. This "one-stop shopping" concept is aimed at lowering import and export transaction costs and reducing obstacles to trade, thus encouraging new entrants from poor countries into global markets. There are 121 Trade Points in more than 80 countries.

Most Trade Points are moving toward *virtual* trade facilitation. A Trade Point client can exchange transaction information with the trade services providers online, either from Trade Point offices or (with the right equipment) from their own premises. Many Trade Points in advanced economies are packaging a suite of free and fee-based information as well as facilitation services, delivered on a Web site. Fully virtual Trade Points include virtual fairs and online malls to present goods and services. They support the entire sales and buy process by extending the Trade Point Electronic Trading Opportunity (ETO) system—basically a large bulletin board on the Web where hundreds of buy and sell requests are matched each day—to include a negotiation and price-finding cycle as well as full user authentication. They also support an Internet-based payment process.

The momentum triggered by the Internet has led governments in developing countries to become more market-minded in spite of uncertainties. But there is still much debate about where and how political and commercial responsibilities need to be separated. These developments present governments with new opportunities but also mean new challenges, as they may alter the traditional relationship with their citizens and with other states. To improve access to the Internet in developing countries, and to afford poor countries similar chances to profit from the e-business revolution, the United Nations is employing Trade Points as a door opener.

All of this innovation and collaboration makes Trade Points a worldwide COIN. The program is a superb example not only of how COINs foster creativity, but also of how COINs can think globally and act locally.

CHAPTER 6

CoINs and Communications Technology

Christianity needed half a millennium to spread across Europe. Some of Leonardo da Vinci's more revolutionary ideas needed hundreds of years until they were recognized. The Fugger and Rothschild dynasties needed multiple generations to come to full power, and the Protestant reformers needed hundreds of years and a Thirty Years War to succeed. What Saint Paul, Leonardo da Vinci, Jakob Fugger, Martin Luther, and Mayer Amschel Rothschild had in common was that not only were they exceptionally creative and innovative, but they were also masters of collaboration and communication. They communicated by:

- Physically meeting people *face-to-face* to build trust and resolve critical issues
- Collaborating with larger groups of people by *broadcasting* what they had to say
- Exchanging *one-to-one* messages with others to learn about new developments and to send out directions

Thanks to the Internet, today's collaborative innovation networks (CoINs) have reached the tipping point where innovative ideas can spread out and literally change the world in days or weeks. From the invention of the printing press to the widespread use of e-mail, information dissemination has gained an unstoppable momentum. Johannes Gutenberg's printing press took 50 years to become widely used across Europe; it then remained basically unchanged until 1874, when a well-known gun maker by the name of Remington brought the first commercial typewriter to the market, allowing for more efficient writing of one-to-one messages. Similarly, in order to speed up delivery of messages, Samuel Morse constructed the first telegraph line from Washington, D.C., to Baltimore in 1844. By the time he died in 1872, telegraph lines connected most of the United States and Europe.

Widening access to sending and receiving information even further, the basic principles of the fax machine were invented in the second half of the 19th century, and the first fax was transmitted from Munich, Germany, to Berlin in 1907. Xerox introduced faxes that would operate over ordinary phone lines in 1967. It took another 20 years for faxes to become ubiquitous in offices, but by then communication was rapidly getting faster and cheaper. Douglas Engelbart first demonstrated e-mail in 1968 at the Fall Joint Computer Conference in San Francisco. Today, e-mail is one of the preferred means of communication in the office and at home, and it is used to link global communities of every sort across continents.

Many-to-Many Multicast

Thanks to the Internet, a COIN can now instantaneously take up, disseminate, and further develop creative ideas, reaching virtually anybody on the globe in an instant and at a negligible transaction cost. Technology has dramatically lowered the cost of all three types of communication depicted in figure 6.1, while at the same time immensely increasing accessibility and speed. Communication technologies have reached the tipping point, where those three means of communication converge into what I call *many-to-many multicast*—meaning that a group of people can start to communicate at exactly the time they want, with whomever they want, and wherever they are.

Even in the age of mass media, however, there is no substitute for direct, interpersonal, *face-to-face contact*. To set up a productive collaboration network, trust between network members must be established. The fastest way to do this is by meeting face-to-face (sidebar 4.4 discusses how to build trust online). This means that people need to travel physically. Today, we can reach any large city within one day, which means that we can initiate and further

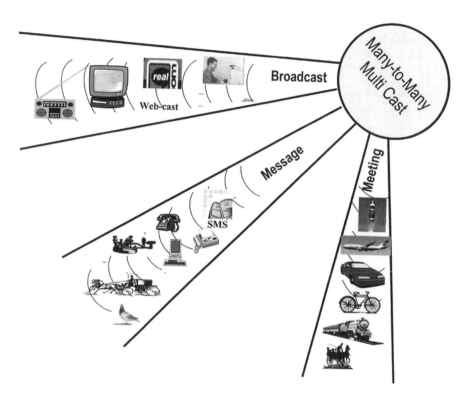

Figure 6.1.   The three communication types at the tipping point.

cultivate COINs at crystallization events such as conferences, in project team meetings, and in working groups; there are even COINs whose seeds have been planted in bed-and-breakfast inns in Cambridge, Massachusetts, on the Cartagena beach, or on a ski lift in the Swiss Alps. The crucial point is that, thanks to vastly increased physical reach, we are now much more likely to meet similarly minded people at such COIN crystallization events.

Once a relationship of sorts among a group of people has been established, they will communicate by exchanging *one-to-one messages*. The successive inventions over the last 150 years of telegraph, fax, and e-mail have sped up delivery and democratized access to one-to-one messaging, and today—as the example of Tom Ojanga in Kenya illustrates—we are capable of reaching anybody on earth almost instantaneously at a transaction cost close to zero.

To build trust on the team level, there must also be a way to recreate the community feeling by reaching groups of individuals at once, namely by *broadcasting* the message. The Internet has put control of broadcasting back into the hands of the audience. Technologies such as Web conferencing and Webcast make highly targeted, many-to-many multicasts possible, allowing individuals to decide whether they want to join a live Web conference or whether they would like to watch a later playback. They can also choose the role they want to assume in the Web conference. Web conferences can even recreate some of the emotional closeness of getting together in a face-to-face meeting[1]—although there is still no real substitute for face-to-face meetings to develop trust quickly.

Three Dimensions of Online Behavior

Model communicators in COINs, such as Tom Ojanga in his parliamentary campaign and Pascal Marmier at SHARE, obey a set of rules for how to interact with each other. Figure 6.2 characterizes the online behavior of individuals participating in online communities along three dimensions: interactivity, connectivity, and sharing. Individuals such as Marmier respond quickly to an e-mail: highly interactive, they know many people with whom they like to connect, and they gladly share what they know with their friends.

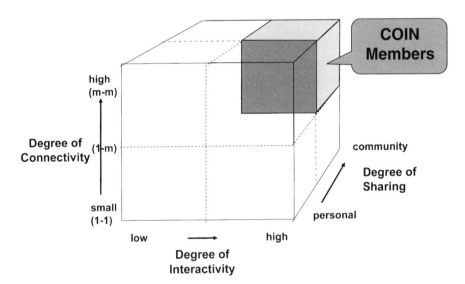

Figure 6.2.   Three dimensions of collaborative behavior: interact, connect, share.

The first dimension of collaborative behavior measures the degree of interactivity. The more interactive online users are, the better for their virtual personalities and the community life of the virtual communities in which they participate. Highly interactive people are responsive, usually replying to e-mails the same day they are received, and they also make themselves available via online chat. This leads to a vibrant community that operates at a fast pace.

The second dimension measures a person's degree of connectivity. Within an online community, team members should be highly connected so that everyone responds to messages from everyone else within the group. A wide and active social network that results from being highly connected is crucial for getting the job done, whatever the job—the more connected people are, the better off the community in which they are members. COIN core team members preferably all know each other and are socially connected.

The third dimension of collaborative behavior measures the degree of group connectivity and knowledge sharing. The more team members are willing to share what they know, the higher the quality of their shared deliverable.[2]

Figure 6.2 illustrates where COIN members fall within the three dimensions of collaborative behavior. The more members of a virtual community interact, connect, and share, the better the online community operates. In a high-functioning COIN, members—be they creators, communicators, collaborators, or knowledge experts—are all connected, interact with each other frequently, and share what they know and find out. To do this virtually, COIN members rely on a collaborative Web workspace.

Collaborative Web Workspace

In collaborative Web workspaces, Web conferencing tools such as Placeware, e-mail, chat, and distributed authoring and editing complement each other and allow virtual teams to communicate and collaborate seamlessly in many-to-many multicast. Placeware is just one example of the more advanced Internet communication and collaboration packages that allow for many-to-many multicasting. Such multicasting systems, combining asynchronous and simultaneous communication, can be built easily with today's Web technologies.

■

Sidebar 6.1
Some COINs Even Speak Their Own Language

Having a common language is a strongly defining cultural element. Today's Western teenagers, for instance, write in their own phonetic language—which has become a natural part of teen culture—when they use chat, e-mail, and mobile phone text messaging (SMS) technologies. Adults have a difficult time understanding.

The British Broadcasting Corporation recently reported on an essay written by a 13-year-old Scottish girl that began as follows: "My smmr hols wr CWOT, B4, we used 2go2 NY 2C my bro, his GF & thr 3 :- kids FTF. ILNY, it's a gr8 plc."[1] Only other members of the SMS culture would understand. The translation is: "My summer holidays were a complete waste of time. Before, we used to go to New York to see my brother, his girlfriend, and their three screaming kids face-to-face. I love New York, it is a great place."

Such cryptic language is a strong defining element of today's web-literate teenagers. To quote Shakespeare: "2b or not 2b that's ?"

Like today's techie teenagers, IETFers have their own language. They list their BCP abbreviations FYI in a FAQ list, inviting to BOFs and asking for RFCs.[2]

Having a common shorthand language not easily understood by outsiders is a hallmark of high-functioning COINs.

1. See http://news.bbc.co.uk/1/hi/uk/2814235.stm.

2. Translation: they list their best current practice abbreviations for your information in a frequently asked question list, inviting birds of a feather to informal workshops and asking for requests for comments (RFCs).

■

Figure 6.3 extends the three dimensions of online behavior introduced in figure 6.2—connectivity, interactivity, and sharing—into a technology framework that combines the pieces of a highly personalized, globally accessible, collaborative Web workspace.

First, the system should offer high *connectivity*. This means the system should include asynchronous, non-interactive tools for one-to-one interaction such as e-mail, but also non-interactive tools for one-to-many interaction such as Weblogs or blogs, where one person publishes individual content for many to read. Finally, it should contain non-interactive high-connectivity tools, such as online bulletin boards, where many people can participate simultaneously in posting and reading messages.

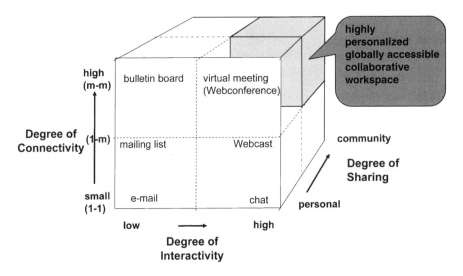

Figure 6.3.   Three dimensions of collaboration technologies.

Second, the system should combine non-interactive tools with tools of high *interactivity*. E-mail should be complemented with chat or instant messaging for interactive one-to-one messaging. Similarly, Webcasts that allow one person to address a group audience interactively can complement and extend the functionality of non-interactive Weblogs. The online bulletin board should be supplemented by interactive Web conferences, where tools such as Placeware permit a group of people to simultaneously work together online by interactively sharing data.

Third, while an individual can use such a system to manage her or his personal knowledge, the system delivers its full potential only if it is utilized by the team to connect as a group to *share knowledge* globally. The result will be a highly personalized, but globally accessible, workspace open to collaboration among all team members. For a community, such a system provides a collective memory to leverage fully the potential of its members. The collaborative Web workspace integrates e-mail, chat, blogs, Webcasts, bulletin boards, and Web conferencing into one seamless system.

My personal experience has shown that a collaborative Web workspace is a useful tool with which an online community can stay in touch and collaborate. But for the success of COINs, the quality of the *people* network is far more important than having the best collaboration tools. The most important prerequisite for a well-functioning COIN is having a collaborative

■

Sidebar 6.2
Different Behavioral Patterns among
Different Types of COIN Members

While all COIN members appear in the upper right quadrant of figure 6.2, creators, collaborators, and communicators do exhibit different behavioral patterns.

Creators are the visionaries who provide the inspiration to the group. They obtain guru status, where their expressions of the group's vision and goals are taken to form the core of the trust and belief system. These COIN leaders excel in sharing credit and inspiring a feeling of togetherness; they also tend to be somewhat less active senders of e-mail messages, while receiving larger numbers of messages from members of the core group. (Frequently they are not the target recipients, but they receive copies.)

In his writing on open source software, Eric Steven Raymond, describing visionary leaders, notes the following:

[I]nterestingly enough, you will quickly find that if you are completely and self-deprecatingly truthful about how much you owe other people, the world at large will treat you as though you did every bit of the invention yourself and are just being becomingly modest about your innate genius. We can all see how well this worked for Linus [Torvalds, the creator of Linux]!'

Communicators or ambassadors sell the vision and the ideas of the COIN to the outside world. They usually combine good connections inside and outside the

culture in place. Absent mutual trust and a willingness to share knowledge, the best collaboration tools are useless.

Outlook for the Future

The process of innovation by COINs happens in these three phases:

1. Invent
2. Create, collaborate, communicate
3. Sell

hosting organization with good "selling" skills, and they know how to formulate the value proposition of the COIN so that it is received well by the managers of the hosting organization. In their communication behavior, communicators send more messages than they receive, with a large fraction of their messages going to recipients outside of the core group.

Collaborators, the most active and committed people within the COIN, coordinate the day-to-day community activities and ensure that the infrastructure is in place to meet the community's objectives. For example, they set up collaborative Web workplaces. They may also assume the role of "friendly policemen," educating novice community members on proper etiquette and behavior. Collaborators are the COIN's most active e-mailers, usually sending more messages than they receive. Most of the recipients of their messages are people in the core group.

The community also includes subject matter experts in areas relevant to the COIN's goals. These knowledge experts operate inside the COIN, and they are mainstays of the community in two ways: (a) they have the deep skills and knowledge to make the community's vision come true, and (b) they work hard to transform the shared vision into a real product. They are also tutors and mentors. Their communication behavior is less active than that of creators, and they primarily receive targeted messages from within the community.

1. Raymond, *The Cathedral and the Bazaar* (1999). See especially version 3.0 at http://catb.org/ēsr/writings/cathedral-bazaar/cathedral-bazaar/ar01s07.html.

The COIN is active in the middle phase, creatively collaborating and communicating to convert an idea or invention into a real innovation. Innovation by COINs happens in organizations that nurture a culture for COINs based on meritocracy, consistency, and transparency. The determinants of this culture for COINs are self-organization and swarm creativity, operation by an ethical code, and communication in small-world networks by hubs of trust. The diffusion of innovation and the creation of new innovations come about through larger networks, and there are tools to optimize the flow of knowledge so that organizations can make ideal use of these online communities. (These are among the subjects of this book's appendixes.)

High-functioning COINs comprise highly motivated creators, collaborators, and communicators. Creators working together in massive collaborative creativity form the core of COINs. Communicators who bridge structural holes in small-world networks are key to more efficient organizations. By acting as gatekeepers, they make possible the optimal redesign of the flow of knowledge. Collaborators, personifying the ethical code of COINs, form the foundation of the COIN.

COINs have been active for centuries. Today, though, thanks to the many-to-many multicast capabilities of the Internet, self-organizing virtual collaborative communities have never been more important. Driven by an environment of high trust and operating in internal transparency, COINs are a dominant driver of innovation. Organizations that foster a culture supportive of COINs can expect an innovative boost. Creating ethical, small-world, "swarming" virtual communities is now a decisive factor for high-performing corporations.

Organizations that explicitly support a culture for COINs have a more efficient operating environment, where self-motivated individuals create and share knowledge. By building a network where core COIN members communicate as hubs of trust in a scale-free, small-world structure, reaction time and responsiveness to change are greatly enhanced.

So the future has just begun! While every innovation can be traced back to individuals, it is collaboration among creators that brings disruptive innovations over the tipping point. The unprecedented capabilities of the Internet in matchmaking and connecting individuals are opening up new avenues to innovation at a higher speed than ever before.

How do you combine all the principles, DNA, elements, technologies, roles and responsibilities, and communication patterns we've explored to create COINs? The following three appendixes detail three methods and tools for reaping the benefits of higher creativity, quality, efficiency, and faster results at lower costs.

Appendixes

In 1968, Internet visionaries J. C. R. Licklider and Robert Taylor described a brave new world "linked" by computers.

In a few years, men will be able to communicate more effectively through a machine than face to face.

What will on-line interactive communities be like?

In most fields they will consist of geographically separated members, sometimes grouped in small clusters and sometimes working individually. They will be communities not of common location, but of common interest.

You will not send a letter or a telegram; you will simply identify the people whose files should be linked to yours and the parts to which they should be linked—and perhaps specify a coefficient of urgency. You will seldom make a telephone call; you will ask the network to link your consoles together.

You will seldom make a purely business trip, because linking consoles will be so much more efficient. When you do visit another person with the object of intellectual communication, you and he will sit at a two-place console and interact as much through it as face to face.[1]

Today, thanks to technologies that are no longer even at the leading edge, Licklider and Taylor's vision has become reality. For a virtual community to work together successfully over long distances, you don't need the latest Internet gadgets. As we saw with my friend Tom Ojanga in chapter 3, for instance, even a parliamentary campaign in a developing country can be conducted over long distance. Ojanga's case is a convincing example of how a fully operational collaborative innovation network (COIN) can be set up and operated using nothing but e-mail accessed from Internet cafés.

The following three appendixes explore how the vision expressed above can be combined with COINs. While it may appear that COINs seem to come to life serendipitously at the initiative of intrinsically motivated individuals without organizational blessing, the good news is there are actual strategies an organization can employ to *uncover, cultivate,* and *nurture* fledgling COINs to become more effective. In fact, there are even things individuals can do to become more productive COIN members.

The appendixes offer advice for how to be a better COIN member, how to support fledgling COINs, and, most important, how to apply COIN principles in conventional organizations. Appendix A shows how organizations can leverage their COINs and combine customers and suppliers into a seamlessly integrated value network by embedding the COINs in a broader ecosystem called a collaborative knowledge network. Appendix B introduces the temporal communication flow analysis (TeCFlow) visualizer software tool, which gives organizations a simple way to discover COINs by analyzing the evolution of individual, departmental, and enterprise-wide communication patterns. Finally, appendix C introduces knowledge flow optimization (KFO) as a means (a) for organizations to reassign COIN tasks and (b) for individuals to become efficient COIN members and fully leverage their skills as creators, collaborators, and communicators.

Collaborative Knowledge Networks (CKNs)

> The next best thing to having good ideas is recognizing good
> ideas from your users. Sometimes the latter is better.
> —*Eric Stephen Raymond*
> *The Cathedral and the Bazaar* (1999)

Three types of virtual communities work together to form an ecosystem of interconnected communities. In chapters 1 through 6, we learned about one type, *collaborative innovation networks* (COINs). The others are:

- *Collaborative interest networks* (CINs), comprising people who share the same interests but do little actual work together in a virtual team. The overwhelming majority of a CIN's population is made up of silent readers or information seekers, called "lurkers" in Internet language, who silently visit a Web site without contributing any content; the minority is a small group of active experts who share what they know with the lurkers.
- *Collaborative learning networks* (CLNs), comprising people who come together in a community and share not only a common interest but also common knowledge and a common practice.[1] People in these networks typically join the community to get to know and learn from like-minded people.

Together, the ecosystem that the virtual communities create is a *collaborative knowledge network* (CKN)—a high-speed feedback loop in which the innovative results of COINs are immediately taken up and tested, refined or rejected by learning and interest networks, and fed back to the originating COINs. The CKN ecosystem is the main mechanism by which COIN innovations are carried over the tipping point.

Table A.1 compares the three types of virtual communities.

The large difference in group size from small COINs to huge CINs is inversely proportional to the focus of the group members. A COIN has the smallest number of members, but they are the most dedicated and focused. A CIN has a far greater number of members, but there is also far less overlap of interests. Typically, a COIN will have a dozen core members at most, whereas a CLN can grow to a significantly larger core size; for example, the community of Xerox repair technicians includes thousands of engineers. CINs are even larger: the Motley Fool claims to have millions of community members looking for, and sharing, investment tips.[2] And a CIN's focus is wider and wider. The Motley Fool's unifying theme of investment tips is much broader than the topic of further development of the semantic Web cultivated by the working group of the same name at the World Wide Web Consortium (W3C).

The CKN ecosystem works as a two-way carrier system of innovation. First, it acts as a dissemination mechanism for the ideas developed by COINs,

Table A.1.   Comparison of COIN, CLN, and CIN.

Community Type	Category	Focus	Mode of Participation	Example
COIN	Innovation	Fundamentally new insights	Peer group of innovators	Linux kernel developers, creators of the Web
CLN	Best-practice knowledge stewarding	Shared knowledge	Active sharers of knowledge as experts; active seekers of knowledge as students	Xerox repair technicians, oneFish, Web masters
CIN	Helping	Shared interest	Few sharers of knowledge as experts, many (silent) seekers of knowledge, lurkers	The Motley Fool, Internet users

Sidebar A.1
Examples of Other Kinds of Networks

All the people participating in the Motley Fool online community (www.motley.com) share a common interest in financial investments. But their collaboration doesn't go very far; all they do is share investment advice, helping each other in a weak form of altruism. People contribute what they know because they believe they will profit from the knowledge of others. But in the Motley Fool community, people do not work together in teams toward a common goal. Instead, they work together because each participant wants to become *independently* wealthy.

The Motley Fool is an ideal example of a CIN. People interested in the future development of the Internet comprise another good CIN example. The huge group of "lurkers" collects information from Web sites such as www.slashdot.org, www.ietf.org, and www.w3c.org, but do not add content.[1]

The oneFish community (www.onefish.org), on the other hand, is a typical CLN in which people share knowledge—in this case, on aquaculture, aquariums, algae cultures, policies and regulations, and shellfish cultures. The community Web site sports an active bulletin board as well as a vast list of resources, including funding sources, United Nations programs, and listings of experts. It includes a virtual office, where interested individuals and organizations can work together virtually on projects. For example, Project Seahorse on the oneFish community Web site links to another site where a team of biologists and social workers collaborate on "conserving and managing seahorses, their relatives and their habitats while respecting human needs."[2]

Another example of a CLN is the group of people—known as "Webmasters"—who develop or maintain Web sites. They are interested in the further development of the Web and they usually share what they know on portals such as www.apache.org[3] and www.slashdot.org, as well as in the discussion groups of the World Wide Web Consortium. Here, the Webmasters are not lurkers but active knowledge seekers.

1. Sponsored by Open Source Technology Group (OSTG), the Internet Engineering Task Force, and the World Wide Web Consortium, respectively.

2. See http://seahorse.fisheries.ubc.ca/.

3. Sponsored by the Apache Software Foundation for open source software developers.

exhibiting a ripple effect of open collaborative innovation. Second, it works as an innovation generator and idea-hatching mechanism, extending into an infinite helix structure of open collaborative innovation.

Innovation Dissemination: The Ripple Effect

COINs are embedded into a multidimensional network of virtual communities and are at the center of a set of concentric communities, in which each community is included within the subsequent, larger community. The dissemination of new ideas is very similar to the ripple effect observed when a pebble drops into water. Innovations ripple from the innermost COIN circle to the next larger CLN circle, and then to the CIN circle, until they reach the rest of the virtual world—eventually moving from the virtual into the real world. Figure A.1 illustrates what I call the "ripple effect of collaborative innovation," using Linux open source developers as an example.

The creative people around Linus Torvalds threw the pebble in the water. Over time, a small group of volunteers, starting out as a COIN, slowly attracted more people on their periphery who were not necessarily interested in becoming members of the core COIN. For instance, the Linux kernel developers around Torvalds soon were surrounded by users who were using

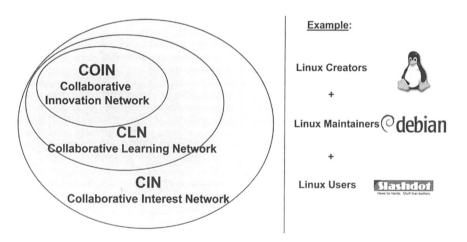

Figure A.1.   CKN-based innovation diffusion by the ripple effect.

Linux to learn about operating system fundamentals, but lacked the skills, time, or interest to join the core Linux development COIN. Nevertheless, they participated in Linux mailing lists or attended Linux conferences.

These people were forming the larger Linux CLN. The most dedicated of these new Linux aficionados formed their own CLN of Linux source code maintainers by, for example, getting together in the Linux Debian community introduced in chapter 3. Core Linux developers and Debian community members found themselves surrounded by Linux users, including Linux system administrators interested in the latest Linux developments, because they make their living through their Linux skills. Later, other people fascinated by Linux joined Linux developers and administrators; this grew into the CIN comprising everyone interested in Linux.

The CLN was extended by end users using Linux simply because it offered a flexible and cost-effective alternative to other operating systems. They are all part of the Linux CKN, getting together on Slashdot.org, the main virtual hangout for Linux users. New COINs grew from the Linux CIN, including one of people dedicated to developing a particular Linux dialect (called a "Linux distribution"). Later, some of these new COINs evolved into commercial companies, including Red Hat, VA Unix, and SuSE. System administrators subscribing to the Red Hat distribution then formed their own CLN; others interested in the Red Hat distribution joined the CLN, and it was extended into a new CIN growing around the Red Hat dialect of Linux.

This biological process of growth and segmentation of virtual communities is not only a major vehicle for the dissemination of innovation, but also forms an *incubator* for new innovations.

Innovative Inspiration: The Double Helix of CKNs

The evolution of Linux and the World Wide Web is a prime example of CKN-based creation of new innovations. The group of early Web innovators and enthusiasts led by Tim Berners-Lee and Robert Cailliau soon attracted people who were not innovators, but active maintainers of Web sites, or "Webmasters." Not all Webmasters became members of the core COIN, but they all practiced the same craft of Web site maintenance, and all were members of the Webmaster CLN. Later, they were joined by non-technical people who had an interest in the development of the Web, such as investors in Web technology

Sidebar A.2
Software Facilitates Networks that Make Up the CKN Ecosystem

Software for CKNs falls broadly into three categories to help locate people, manage content, and facilitate working together.

Software tools in the first category uncover and explore hidden links among people. They can be used for a variety of purposes, from dating to finding people with similar interests to developing business networks. Friendster (www.friendster.com), for example, is an online community that connects people through networks of existing friends for dating or making new friends. Tribe.net broadens the focus to any type of interest community, inviting people to get

Three categories of CKN support software

	Locate members	Manage contents	Collaborate
Interest (CIN)	Find people with same interests (Friendster, LinkedIn, Spoke, Yahoo groups)	Web sites, Web logs	e-mail, chat, mailing lists
Learning (CLN)	Find tutors (Tribe.net, elance.com)	Management of learning content (skillsoft)	Real-time online courses (Placeware, Webex, Saba)
Innovation (COIN)	Find experts (Google Answers, wer-weiss-was.de)	Knowledge management (Autonomy)	Collaborative Web Workspace (Groove, Lotus Workplace, LiveLink, Intraspect, PTC)

stock or Internet journalists. These people did not innovate, and they did not practice the craft of Web site maintenance, but they shared an interest in the Web's further development and were thus members of the Web CIN.

The knowledge in these communities disseminates from one type of community to the next. Over time, a COIN extends into a CLN, which then expands into a wider CIN. The process does not stop once a CLN has broadened into a CIN. Rather, new innovative ideas arising in the CIN will be taken up by new COINs, as figure A.2 illustrates. For example, the original innovators of the Web have taken up the concept of the "Semantic Web," where information is described by content-based metadata. They are

together and join "tribes" whenever they need a new apartment or a new job, want to know about the coolest bars and best restaurants, need to buy used furniture, or just want to make new friends. Several software vendors offer similar services to tap into networking for business contacts.[1]

Other software tools manage unstructured knowledge, making it possible to store and retrieve unstructured data such as text, images, video, and audio. They also have sophisticated filtering, visualization, and content mapping capabilities. Often, they can be personalized, and they offer search and retrieval agents that notify the user when relevant information has arrived.[2]

The transformation of work processes by Internet-based communication tools and technologies has just begun, and no software vendor yet offers a fully integrated collaborative Web workplace (although a number of partial solutions are available).[3] Looking back at the growing speed of change in communication technologies culminating in the many-to-many multicast introduced in chapter 6, it's safe to predict a few years from now we will be using such tools in ways unimaginable today.

1. Friendster, Spoke, ZeroDegrees, Ryze, and LinkedIn are both software tools *and* collaboration portal vendors. The distinction is a fluid one (and can be confusing), because they all have custom-built software usually not for sale, but their business models are all based on having the largest online community network—that is, a portal. Analysts refer to them as social "networking software vendors," which may well be the best name.

2. Vendors include Autonomy, OpenText, Eurospider, Engenium, and Entopia.

3. Lotus Notes pioneered these concepts more than 10 years ago, and Notes-based applications are still among the leading collaboration software packages. Other vendors include Groove, which offers a peer-to-peer solution without a centralized server; eRoom/Documentum (now part of EMC), which offers a server-based solution; and Intraspect/Vignette, which is also server-based. Smaller vendors include Tomoye, Webex, iManage, Metalayer, and Stratify, to name just a few.

convinced that this will lead to a better Web, where it will be much easier to locate information. A new COIN has developed around the Semantic Web, which has already expanded into a new CLN. And there are dozens of other COINs active under the auspices of W3C, exploring further innovative extensions of the Web.

Figure A.3 shows how the upward spiral of dissemination of innovation shown in figure A.2 can be extended into a double helix of innovation by CKNs—a potentially perpetual source of new innovations.

In a COIN, the main activity is to create new things; in a CLN, it is to collaborate, to learn about the innovation and understand how to use it; and

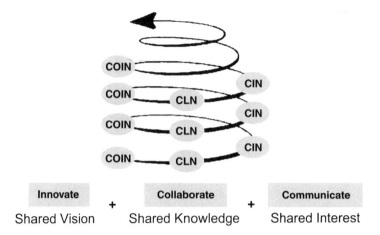

Innovate + **Collaborate** + **Communicate**

Shared Vision Shared Knowledge Shared Interest

Figure A.2. The helix of collaborative innovation. Dissemination of innovation from COIN to CLN to CIN to new COINs.

in a CIN, members communicate, carrying news about the innovation to the rest of the world. In figure A.3, one helix displays the evolution of the virtual communities—a COIN spawns a CLN, which extends to a CIN, which in turns spawns new COINs, which grow into new CLNs, and so on. The other helix illustrates the main activities of the individual CKN members, who progress from innovating to collaborating to communicating and then again to innovating, and so on. Sidebar A.3 offers a real-life example.

The Deloitte e.Xpert Practice as a CKN

The pattern of growth and expansion of the Deloitte Consulting e.Xpert practice detailed in chapter 5 displays all the characteristics of a CKN at work. In late 1998, the firm's e-business leadership team, led by Deloitte's global e-business practice leader, Cathy Benko, got together to create a vehicle for internal and external expansion of e-business consulting capabilities. Three senior Deloitte partners formed an e-business council to kick off our e-business initiative in Europe, and they worked with me to create the e.Xperts. As the four of us spread out across Europe to win support for our strategy from the managing directors of the different country organizations, we also put together a team of consultants to do all the legwork, composed of a number

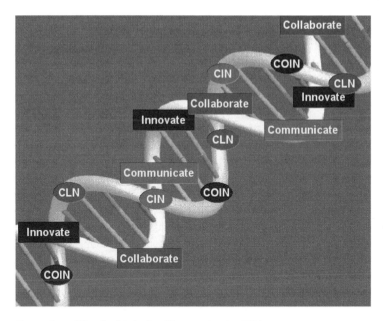

Figure A.3.   The double helix of innovation by CKN.

of highly motivated junior members of Deloitte country offices in London, Zurich, Brussels, Hanover, and Paris. They became the e.xpert core team and did a stellar job getting the virtual pan-European e.xpert practice up and running.

The e-business council and e.xpert core team made up the nucleus of the e.xpert COIN, soon reinforced by the e-Champions—e-business leaders in the 14 country organizations and different Deloitte service lines. Together, we created a training and e-certification program and held a series of face-to-face and virtual conferences in which all the e.Xperts participated. The entire 200 people e.xpert practice formed the CLN of e-business experts learning from each other, while Deloitte Europe's 3000 consultants were part of the surrounding e.xpert CIN. Frequently, members of the e.xpert CIN would graduate to become members of the e.Xpert CLN, ultimately entering the e.xpert COIN as e-champions or members of the e.Xpert core (see figure 5.2 in chapter 5).

The e-Champions leading the teams in the 14 countries—and, even more, those leading e-business teams in the cross-national service lines—created new COINs in their own consulting practices. For example, Finland's

■

Sidebar A.3
CKN = COINs + CLNs + CINs

The Internet Engineering Task Force (IETF) working groups first mentioned in chapter 1 offer another example of CKNs made up of COINs, CLNs, and CINs. The core IETF members of a working group operate as a COIN to develop new Internet standards such as transmission control protocol/Internet protocol (TCP/IP), simple network management protocol (SNMP), or an Internet security protocol.[1] As an IETF standard matures, the COIN working group grows into a CLN of people implementing that particular standard. The CLN members might be working for Internet networking vendors such as CISCO, Nortel, or Lucent. With growing acceptance of a particular IETF Internet standard, the CLN then extends into a wider CIN of early users of new Internet networking equipment built by those vendors. In this larger CIN interest group, small teams of innovators get together to form new COIN working groups to develop new or improved versions of an Internet protocol standard.

IETF has institutionalized this process to the point where new topics can be defined and a new working group can be established as soon as a dedicated team is willing to tackle a new topic. The initial "kickoff" group usually involves three to seven people, and a fully operating working group has between 120 and 170 active members.

Watch many different COINs, CLNs, and CINs in action and a consistent pattern of growth, active membership, and segmentation emerges. A COIN's core group comprises between five and 12 members, and a CLN typically includes 20 to 200 active COIN and CLN members. If a CLN becomes larger, it splits into new COINs, which again exhibit the same structure. In principle, this model runs infinitely and applies not only to the evolution and propagation of the Web and the Internet, and to Linux and open source development, but to all open innovation driven by COINs.

CKNs form an extremely powerful engine of open and disruptive innovation for knowledge creation and dissemination.

1. TCP/IP is the suite of communication protocols used to connect hosts on the Internet. SNMP is a set of protocols for managing complex networks.

■

e-Champion became an early adopter of the CKN idea and built up a virtual team of consultants experienced in the cross-section of e-business and knowledge management consulting, offering CKN-based services to his clients.[3]

The Deloitte e.xpert example shows the ripple effect and the double helix: the CKN concept is the product of a COIN, while also delivering a great showcase for the birth of a COIN and its expansion into a CKN ecosystem.

Conception of the "CKN Idea"—Created by a COIN!

Two colleagues and I were attending a Deloitte Consulting global e-business leadership meeting at a Silicon Valley hotel when the head of Deloitte Research asked whether we would write a research paper for "e-views." We gladly agreed and proposed "Coin" as our topic. At the time, we were using this acronym to signify a "Community of Interest Network."

We began a dialogue with Robin Athey at Deloitte Research, who was working on that very subject. She and I exchanged e-mails, and then arrangements were made for us to meet in New York City. Robin had rounded up a team of seven Deloitte consultants, complemented by Thomas W. Malone of the Massachusetts Institute of Technology (MIT) Sloan School of Business, a world-renowned authority in the field of computer-supported collaborative work. Over two days at this first meeting, we discussed what makes up a "Coin." After this face-to-face meeting, Athey kept the ball rolling, polling and pushing us for ideas. We continued to discuss, adding to the topics the scientific ground we wished to cover. Should our focus be on "communities of practice" or on virtual teams collaborating over the Web? I proposed to rename the virtual teams we wanted to study, changing "Coins" to "collaborative knowledge networks," or CKNs. Robin gave it some deep thought, then bought into the name change with enthusiasm, convincing me that this would be the best name for our research report. After another six months of close virtual collaboration, we published our original report on CKN.[4]

Once the report was published, the idea gained traction around the globe, and Deloitte decided to develop a new consulting service based on the CKN concept. The composition of the original team changed considerably over the next 12 months. We gained enthusiastic supporters in Dallas, San Francisco, New Jersey, Brussels, Hanover, Melbourne, Singapore, and elsewhere. The team worked together over long distance as an ecosystem of COINs (collaborative innovation networks), extending into a CKN with hundreds of participants. Different teams were born, working on topics such as an advanced Web site (which we called "Personal Knowledge Marketplace"), a tool-supported CKN diagnostic to assess the CKN maturity of an organization, and a CKN community-building methodology. Each team operated as a true COIN, wherein core team members acted as self-selected and self-organizing innovators, collaborators, and communicators to the best of their capabilities. They did most of this work in their "spare time" when not fulfilling their responsibilities as consultants on client projects. Most of the work was done wherever the consultants were located, through virtual cyberteamworking.

Every Thursday at 4 p.m. local time in Switzerland (the most convenient time we could come up with given global participation), we had a virtual meeting of the COIN on "CKN" using the Placeware Web conferencing system and a global conference call. We would discuss work in progress and the latest results, and sometimes we would invite external experts to present their work. The team growing around the "CKN Brown Bag" meeting was structured as a real CKN: the COIN core team was surrounded by a CLN that regularly and actively participated in the meetings, learning about new results on "CKN" and themselves becoming "CKN experts." They were joined by a more peripherally interested group of people who formed a surrounding CIN. Some of the more interested CIN members graduated into the CLN and eventually became COIN core team members.

We also organized a large virtual CKN conference using real-time, Web-based conferencing software, educating the rest of Deloitte about the goals and opportunities of the CKN initiative. Speakers from locations across the United States and Europe addressed a truly global audience (which meant, of course, that some participants had to get up in the middle of the night to hear the presentations live). This conference was a big success and, because of heavy oversubscription, we had to do reruns for local target audiences. This ensured a convenient time for the audience, but now it was the speakers who had to deliver their presentations in the middle of the night.

The CKN core team worked together closely and successfully for almost a year before meeting face-to-face for the first time. That first face-to-face meeting greatly boosted team spirit and morale, but we had already built up so much distributed trust that we cooperated efficiently even prior to the personal meeting. The main reason for this, in my view, was the cohesive culture of Deloitte Consulting, which puts strong emphasis on people-centered values and is less hierarchical and more egalitarian than the cultures of many other management consulting firms. The CKN team was, therefore, able to collaborate in a climate of confidence, based on recommendations of hubs of trust, which acted as mutually trusted gatekeepers and bridged the "structural holes."[5]

What Ends a COIN?

COINs do not last forever; rather, they continuously reinvent themselves in the ways discussed earlier in this appendix. There is a natural growth path: a COIN extends into a CLN; this grows into a CIN that carries a germ of innovation that gives rise to new COINs from its different subcommunities. Each CIN has a small team of hard-core volunteers who band together in new COINs to develop new ideas in their shared fields of interest. These fledgling COINs then grow again to become CLNs and CINs.

Something else can happen, too. Truly successful COINs, especially ones in which the innovations have commercial potential, transform into startup companies, project teams, and business units.

As soon as Union Bank of Switzerland (UBS) realized the value of its COINs working on the Bank Wide Web and the Interoperability Service Interface (ISI; see chapter 5), they were converted into officially sanctioned project teams. This raised the profile of both projects and ensured the allocation of official resources (people and money). Once it was clear that the Web and ISI concepts were there to stay, the project teams were converted into business units, giving them even greater official standing.

Thus, the bank's COINs converted gradually from free-floating, self-organizing groups to highly structured, hierarchical organizations. Figure A.4 illustrates this transformation from COIN to CLN to CIN and from COIN to project team to business unit.

Figure A.4 also illustrates the change over time in the different cohesion factors or unifying themes of COINs, CLNs, CINs, project teams, and business units. The motivating factor triggering participation in these organizations changes from sharing a common vision to making money. While membership in a COIN is clearly driven by shared vision, participation in a business unit comes by sharing the same activities. Monetary interest, at least in the very beginning, is of secondary concern: Web enthusiasts joined Tim Berners-Lee not for riches, but because they all had the same vision of a shared knowledge

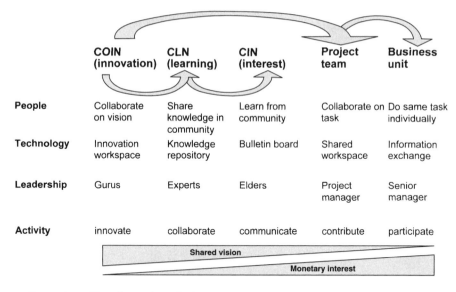

Figure A.4. Transformation of COINs.

universe. The same was true for the first Linux developers who joined Linus Torvalds. It was even true for my colleagues at UBS, as we worked together to get the bank intranet off the ground.

Over time, shared vision can also pay off financially. At the bank, our reward came pretty quickly because our COIN was converted into an official business unit, with group members promoted to its leadership roles. Tim Berners-Lee and Linus Torvalds have won external recognition: Berners-Lee was knighted by Queen Elizabeth II of England and holds the 3Com Founders Chair in Computer Science at MIT; Torvalds, until recently a staff member at Transmeta Corporation in California, has joined the nonprofit Open Source Development Lab as its first fellow.

Of course, all this discussion about networks ought to be raising the big question of how to find the *hidden* COINs, CLNs, CINs, and CKNs in an organization. Appendix B, which follows, introduces a software tool that spots emergent communities by analyzing their communication flow.

Temporal Communication Flow Analysis (TeCFlow)

Imagine listening to a piano sonata by Beethoven. The music is introduced in the exposition and then undergoes a modulation, arriving at the exposition's end in a different key. As the melody develops further, it becomes possible to unravel and analyze the tune. The more time you spend listening to the melody, the better you will understand the underlying pattern.

The temporal communication flow analysis (TeCFlow) tool does for social networks what repeated listening to a Beethoven sonata does for the classical music lover. TeCFlow reveals the evolution of interaction patterns in social networks.[1] Its goal is to build an environment for the visual identification and analysis of the dynamics of communication in social spaces. Just as sheet music defines the way in which the right and left hand of the pianist play together to produce the Beethoven sonata, or as a score defines how an orchestra will play together to create a symphony, the communication patterns of a team define and indicate how that team works together. TeCFlow creates an interactive movie (based on e-mail traffic) that shows this interaction among members of a work team. By comparing dynamic interaction patterns with the performance of virtual teams, TeCFlow can identify typical communication patterns of different types of virtual communities and shed light on how a team collaborates. Along with knowledge flow optimization (described in appendix C), the insights gained through TeCFlow can be used to develop recommendations for improved team performance.

TeCFlow takes as input a communications log and automatically generates interactive movies of the interaction patterns among the involved individuals. Within the movies, each person is represented as a dot. A line between two people indicates a relationship. The closer two people are placed together, the more intensive their relationship (i.e., the more messages they have exchanged). The most active communicators—those who sent and received the most messages—are placed visually in the center of the network. Although TeCFlow is optimized to work with e-mail archives, it can also process chat archives, phone logs, and transcripts of face-to-face interactions. By using this tool, we can uncover hidden collaborative knowledge networks (CKNs) and their component collaborative innovation networks (COINs), collaborative learning networks (CLNs), and collaborative interest networks (CINs).

Note that this is the most technical section of the book thus far, outlining the technical and mathematical underpinnings of the TeCFlow tool. Readers not interested in the technical details of TeCFlow may safely skip to appendix C.

Discovering COINs, CLNs, and CINs with TeCFlow

While all CKN networks display a small-world structure, COINs have the strongest small-world properties, with a degree of separation of only one or two (i.e., two people can always reach each other by contacting a mutual friend). CLNs have a somewhat larger degree of separation of two to three, while CINs have the largest degrees of separation, up to four to six. By using these properties and searching for clusters in TeCFlow-produced movies, we can automatically uncover hidden COINs, CLNs, and CINs.

Figures B.1, B.2, and B.3 use the Deloitte Consulting e.Xpert case (see chapter 5) to illustrate the communication patterns of a typical COIN, CLN, and CIN as visualized by TeCFlow. In figure B.1, wee see the communication network of a COIN working to develop a new service offering for the Deloitte e-business practice. The larger picture shows the entire conversation of the core team, including subject-related communication with people outside the core team; the dense cluster in the center at left is the core team. The smaller inset picture illustrates communication among the COIN core team members, depicting an almost completely connected network in which everyone talks to everyone else directly.

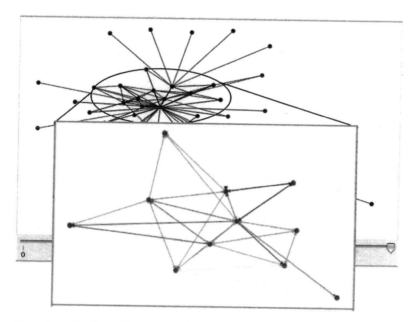

Figure B.1.   Small-world structure of a COIN.

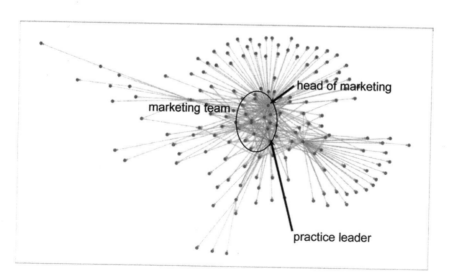

Figure B.2.   Network structure of a CLN.

Figure B.2 displays a snapshot of the TeCFlow movie of Deloitte's e-business practice marketing team, which obtains information from the e-business leadership team and operates as a learning community. The dark dots are members of the e.Xpert marketing team. While the practice leader and the head of marketing are the centers of their own extensive communication networks, the entire marketing team is engaged in mutually dense communication, learning from each other while organizing marketing events for the e-business practice.

Figure B.3 shows communication of the entire e -business practice over a full month. Different subcommunities are clearly recognizable, such as star-shaped learning networks (or CLNs) and highly connected COINs.

Taken together, figures B.1, B.2, and B.3 also illustrate nicely the small-world, scale-free effect explained in chapter 4; they show that the more "small world" and scale free a community is, the more robust it will be, and thus the less likely it will be to disintegrate if key members leave. For instance, a COIN-like structure with a high connectivity among team members in a "small world" is more robust than a CLN. If too many of the subject matter experts who form a CLN's core team leave (figure B.2), it could dissolve. A CIN (figure B.3) is even more prone to falling apart, as it is linked together by

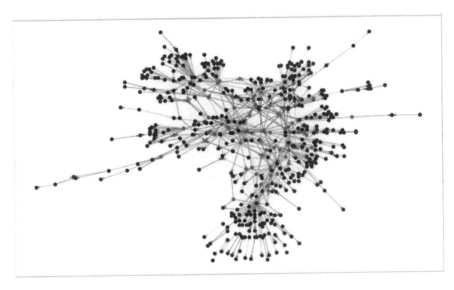

Figure B.3.   Small-world structure of a CIN.

a few people acting as gatekeepers who bridge the "structural holes." If those gatekeepers leave, the CIN will disintegrate into different, smaller interest networks.

By looking at the position of individuals in these dynamic networks, we can gain a better understanding of typical communication patterns of individuals in COINs.

Analyzing Roles and Communication Patterns in COINs with TeCFlow

Open source projects provide an excellent blueprint for defining the various roles in COINs. Our Massachusetts Institute of Technology (MIT) research project has explored the communication behavior of numerous open source communities. As those communities store their exchange of e-mail in publicly accessible mailing list archives, they provide a unique opportunity to analyze how community members interact.

Figure B.4 is a snapshot view—produced using TeCFlow—of the communication pattern of a typical COIN, in this case a working group of the

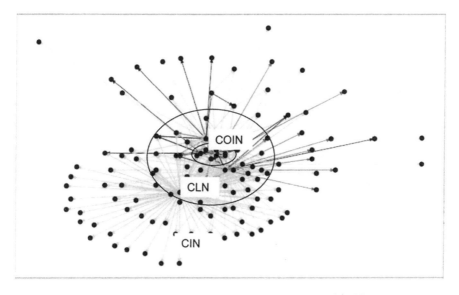

Figure B.4.   E-mail communication in an open source COIN, with COIN core group, surrounded by CLN and CIN.

World Wide Web Consortium (W3C) over nine months. The more e-mail messages two people exchange, the closer they are located to each other in the figure. People sending or receiving the highest total number of messages are in the center and comprise the core group, whose members are, to a large extent, communicating among themselves. This figure shows the three-ring structure of the COIN ripple effect, with an inner core of about five to seven people (see sidebar B.1), a middle-circle CLN of 15 to 20 people who are still quite well networked into the core group, and an affiliated CIN of people who are only connected to one or two core team members.

Our research has analyzed the communication behavior of the core participants of a COIN by looking at how many e-mail messages team members sent and received.[2] We also determined whether the recipients of these messages were other core team members or people outside the core team. We defined what we call the "contribution index" to measure the level of active participation of an individual in the community.

$$\frac{\text{messages sent} - \text{messages received}}{\text{messages sent} + \text{messages received}}$$

Sidebar B.1
COIN Core Groups

The core group is the nucleus of a COIN. Its main activities are to develop new ideas, recognize new ideas from outside for inclusion into the COIN's activities, and "sell" the COIN's innovative ideas to the outside world.

Typically, the core group consists of three to seven people who are totally dedicated to the success of their new idea, usually drawn from different units within the organization hosting (usually unknowingly) the COIN. The core is a mix of senior knowledge experts and enthusiastic junior members; these members forms an initial team that works together to get the COIN off the ground. Once the community is established, the core group typically grows to 10 to 15 members and continues to remain the new community's organizational mainstay.

Membership in the core group will change over the lifetime of the community. Founding members leave because their interests change or for external reasons. Meanwhile, more junior members of the community build up their skills and reputations over time and become accepted members of the core group.

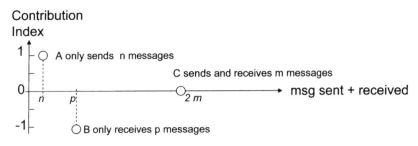

Figure B.5.   Contribution index.

If an individual only sends message and receives no messages, her or his contribution index is $+1$. It is -1 if he or she only receives messages and never sends a message. If the communication behavior is totally balanced—that is, the number of messages sent and received is the same—the contribution index is 0.

We plotted the contribution index against the total number of messages sent and received by each participant in the W3C project. Figure B.5 shows that person A sent n messages and received none; person B received p messages and sent none; and person C sent and received exactly m messages (person C is located on the X-axis because he or she sent and received the same number of messages).

After looking at a large number of innovation networks, we identified four different role patterns (corresponding to the behaviors discussed in chapter 6) for *creators* (gurus), *communicators* (ambassadors), *collaborators* (expediters), and *knowledge experts*. Figure B.6 shows these patterns; the COIN leaders emerge by examining the group communication pattern from e-mail traffic (as described above). Figure B.7 displays the communication pattern of nine appointed leaders (rectangles) of a W3C working group.[3] Not surprisingly, the leaders show up in the center of the figure, along with other significant contributors. People from one organization are more active leaders with more central positions (dark rectangles) than people from the other organization (light rectangles).

Notably, there is a significant difference within the leadership group in contribution frequency (the numbers of messages sent) and in the contribution index (defined above). Figure B.8 illustrates activities of the nine appointed leaders. It shows that people from one organization are more active

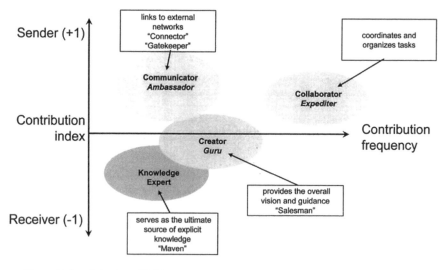

Figure B.6.   Roles in a COIN.

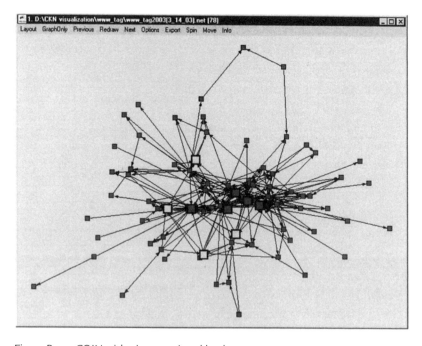

Figure B.7.   COIN with nine appointed leaders.

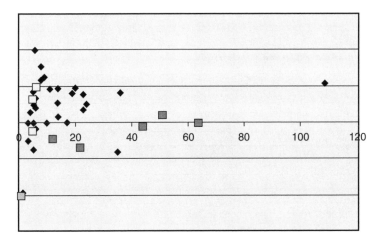

Figure B.8.   Contribution index of COIN members.

contributors (dark rectangles), while people from the other organization are less active (light rectangles). This is true in absolute numbers of messages sent as well as in balanced communication behavior, with a contribution index close to 0, which means the active leaders send and receive approximately the same number of messages. The representatives of the less active organization participate much less in absolute numbers, but some of the members send substantially more messages than they receive. There are also some extremely active participants in the discussion who are not appointed leaders.

Applying TeCFlow to Analyze the Flow of Knowledge in Social Networks

To formally describe the evolution of typical communication patterns in our research, we have been applying results from the field of social network analysis—a subject of active research in the social sciences that correlates statistical properties of the graph structure of social networks with the behavior of the people and the outcome of the work conducted in those networks (see sidebar B.2). In particular, we are calculating the evolution over a time interval of a variable that describes the connectivity structure of the network. This variable is called *group betweenness centrality* (GBC). The lower the GBC, and the higher the graph density at a particular time, the more people

■

Sidebar B.2
COINs in Social Network Analysis

In social network analysis terms, COINs consist of a central cluster of people, the core team, forming a network with low betweenness centrality but high density. Each person in a network has an associated betweenness centrality, which is high if someone is in a central position between many other people.

The GBC of the entire group is 1 for a perfect star structure, where one central person—the star—dominates the communication. GBC is 0 in a totally democratic structure where everyone displays an identical communication pattern.[1]

The external part of a COIN consists of a network forming a ring around the core team. The COIN core team has comparatively high density but low GBC. A CLN or CIN has lower density but higher GBC, as external members are connected only to core team members but not among themselves.

	COIN	CLN/CIN
Core/ periphery	large core small periphery	small core large periphery
Density	high	low
Betweenness centrality	low	high

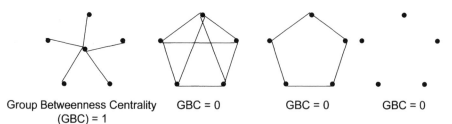

Group Betweenness Centrality (GBC) = 1   GBC = 0   GBC = 0   GBC = 0

GBC varies.

1. For a precise definition of betweenness centrality, see Wasserman and Faust, *Social Network Analysis: Methods and Applications* (1994).

■

in the network are talking to each other "in a democratic way" on even terms. This is a strong indicator for the emergence of a COIN.

A three-step process using TeCFlow reveals the typical communication patterns of virtual communities.

1. Watch TeCFlow communication flow movies to find dense clusters indicating the potential emergence of COINs.
2. Look for peaks and troughs in the temporal evolution of GBC and density to find the most "interesting" phases of collaboration in the team's lifetime.
3. Look at the *contribution index* to understand better the roles of individuals in teams.

TeCFlow offers a novel way to look inside team communication and uncover, with relative ease, information that is very hard to get using other, more conventional means. And by using the e-mail archives of the Deloitte e.Xpert practice, we can illustrate the three-step analytic process.

1. *Watch movies to find COINs.* By watching a movie containing the entire e-mail traffic of the e.Xpert consulting practice and looking for highly connected clusters of people, we reveal potential COINs (see figure B.9). This is a strong indicator for the emergence of an innovation team.
2. *Find interesting time periods.* We then look at the progression of GBC and group density over time. A rapid change in the slope of the graph (i.e., a peak or a trough) indicates an interesting event that warrants going back to the movie and drilling down into the network graph by clicking on interesting people and looking at the headings and contents of the e-mail messages exchanged. The two troughs in figure B.10 correspond to periods of high activity in the COIN.
3. *Identify the most active participants.* Finally, we plot the contribution index to identify the most active people. In our example (figure B.11), we find that the practice coordinator sends the most messages (more than he receives), while the practice leader sends fewer messages and receives significantly more than he sends. We found this to be a typical behavior pattern for group leaders in the online communities we analyzed. Typically, the practice coordinator is also the most

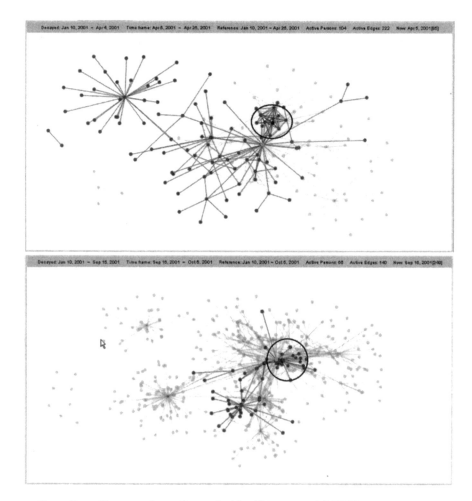

Figure B.9.   Two snapshots of a movie, identifying potential COINs.

active contributor; surprisingly, though, it is a practice *member* in this community who is the most active participant, making herself the leader at the core of a new innovation community.

We can then combine the steps to analyze the birth of a new COIN. The analysis with TeCFlow affords us an intimate look into the emergence of new teams and online communities that would be hard to obtain by other means— had we not combined the contribution index plot with the dynamic movie, the creation of the COIN as well as the emergent role of its leader would

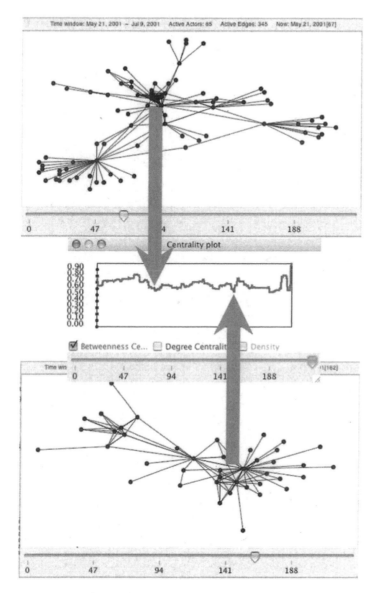

Figure B.10.   Abrupt changes in GBC (middle line).

Figure B.ii.
Contribution
index of e.Xpert
practice
members.

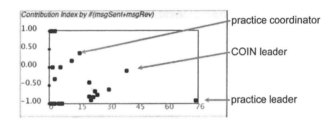

have remained hidden (figure B.12). The GBC view allows us to fast-forward quickly to periods of particular interest, where we then can use the drill-down features of the dynamic view to look at the contents of the e-mail exchange. Our analysis proved the points about COINs made in earlier chapters.

- We saw the emergence of a new innovation team, coming up with a creative new consulting service offering.
- A non-executive member of the consulting practice played the central role in creating this new service offering.
- The period when the new innovation team was most active was easy to identify, and it was also easy to identify the core members of the new innovation team.

This analysis was done two years after the data had been collected. Had the tool been available in real time to monitor virtual interaction, senior management would have been afforded the opportunity to provide adequate support for the new COIN. That, in turn, would have reduced time to market for the innovation and increased awareness of the team's output within the consulting firm. In short, not being aware at the time of what was going on likely cost Deloitte consulting revenue.

Let's consider some specific examples of how TeCFlow analyzes different patterns of knowledge in different situations.

Innovation Pattern: Creation of a New Service Offering

The following example illustrates the interaction pattern of an innovation community. We are analyzing a subset of the Deloitte e.Xpert messages used for the analysis in the previous section of this appendix. The subset was

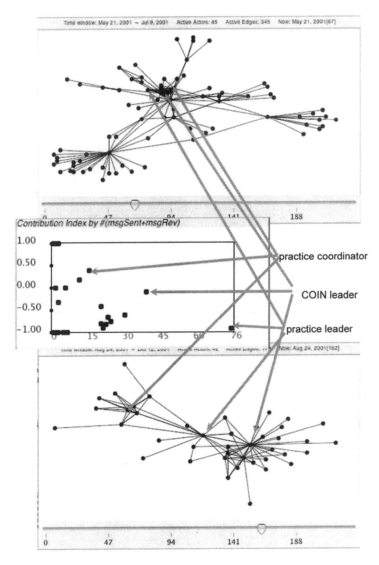

Figure B.12.   Visualization of an entire e.Xpert practice.

categorized manually to include all communication about the creation of the new "CKN diagnostic" service offering, which resulted from the innovative work of the COIN identified above.

A dedicated team of 39 volunteers from eight countries around the globe worked together to create this new consulting service offering. The core team members came from Switzerland, the United Kingdom, the United States, Belgium, and Germany. They never met face-to-face, but they worked together over a six-month period to create a software platform and marketing material that were widely used to win consulting engagements over the next three years.

The first TeCFlow movie snapshot in the upper left of figure B.13 illustrates the low GBC and high-density structure of the innovation team (or COIN). The trough in the GBC plot in the lower left of figure B.13 shows the initial phase where the team collaborated to create the offering, while the peak toward the end of the GBC plot demonstrates the structure with a tightly connected core surrounded by weakly connected people (i.e., the innovation team was encircled by a learning and interest community). This communication pattern is shown in the movie snapshot at the upper right of figure B.13.

The contribution index plot in the lower right of figure B.13 demonstrates the immense efforts of the practice coordinator, who was vital for the project's success; this contribution was not apparent at the time the service offering was created. There is a huge activity gap between the practice coordinator and the rest of the project team. A summer intern hired to help create this offering was only peripherally integrated into the core team, as the analysis shows, because team members did not adequately include her in the communication flow of the project.

Learning Pattern: Information Dissemination by Webinar

The next example illustrates how new concepts are taught to an online audience forming a CLN. Our data set for this example consists of an e-mail archive of the e.Xpert consulting practice, where a team was organizing a global Web-based seminar ("Webinar"). The archive covers a period of about 200 days.

A small team of self-selected members of the e.Xpert practice prepared the Webinar over several months. One main speaker then delivered the Webinar over one hour, assisted by his team members. The audience, spread out globally, had the opportunity to question the speakers via e-mail during the Webinar. Questions that could not be answered during the talk were replied to over the next few days, and because of overwhelming demand, the team decided to rerun the Webinar a few weeks later after making some minor changes. The rerun had a different main speaker.

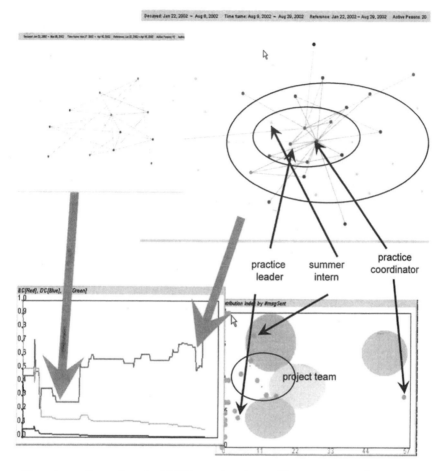

Figure B.13.   Expansion of a COIN into a learning and interest community.

The TeCFlow movie illustrates nicely the switch between an innovation community pattern and a learning community pattern. In the Webinar preparation phase, the core team collaborated as a COIN, whereas in the delivery phase the speakers and audience collaborated as a CLN.

Figure B.14 illustrates the progress of the communication activities over time related to preparing and conducting the Webinar. The center of the figure is a time series plot of the evolution of the GBC and density of organizers and audience during the entire period. The picture at the top left shows the structure of the team preparing the Webinar, which is operating as a COIN, with high

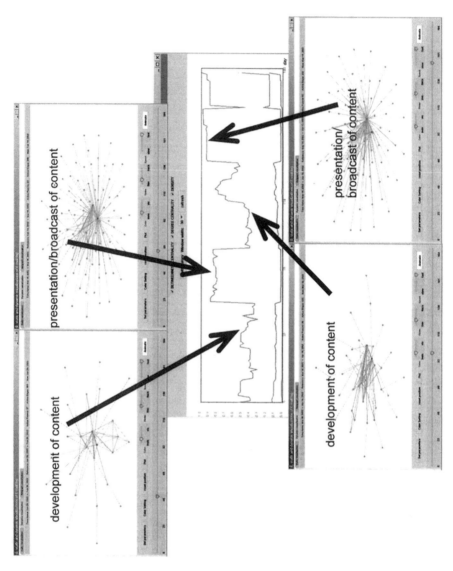

Figure B.14 Webinar visualization with a plot of CBC in the center.

density and low GBC (arrow 1). The picture at the top right is a screen shot of the communication pattern during the first delivery of the Webinar. The practice leader (dark central dot) is sending and receiving information in a star structure; the dent in the graph in the center, as pointed out by arrow 2, identifies this phase of high GBC. During and after this first run of the Webinar, it is the practice leader who fields most questions. The picture at the lower left displays the team preparing a rerun of the Webinar, again working as a COIN and communicating with low GBC (arrow 3). Finally, the screen shot in the lower right displays the practice coordinator (dark central dot) rerunning the Webinar, communicating with his audience and answering their questions in a star structure with high GBC centrality (arrow 4).

Innovation Pattern: Rise and Demise of Startup Cybermap Systems

In sidebar 5.2, Cybermap Systems was introduced as an example of a COIN that failed. Here, with TeCFlow, we can use the Cybermap example to illustrate changes in knowledge flow during the three main phases in the life cycle of a COIN. In this case, TeCFlow was employed to examine the e-mail archive of the startup's founder to visualize and analyze the knowledge flow and communication patterns among team members. The founder's e-mail archive served as a substitute for the organizational memory of the COIN.

Figure B.15 shows the changes in GBC and density during the critical periods in the life of the startup. The GBC plot identifies the three main phases during Cybermap Systems' existence:

1. *Connect:* The core group was connected by the founder.
2. *Collaborate:* The core group worked together as a team.
3. *Dissolve:* The communication flow stalled as the founder was distracted and withdrew from communication with the rest of the team.

Figure B.16 consists of three snapshots of the communication flow movie automatically generated by TeCFlow. The snapshots again illustrate the critical phases in the startup company's life cycle. In picture 1, we see the initial "connect" phase, where the founder (the dot in the center) is connecting the lead programmer (dot close to the founder) and other future core team members. The founder is contacting his various friends, trying to convince

Figure B.15.   GBC and density of Cybermap Systems communication flow.

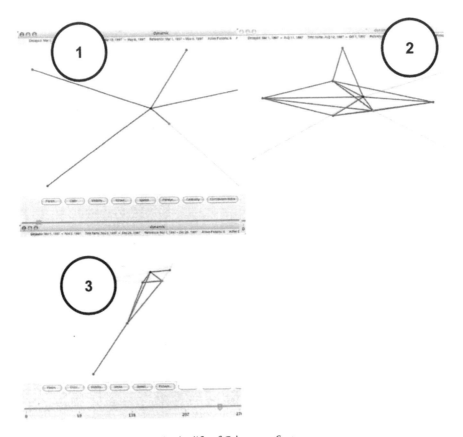

Figure B.16.   Three stages in the life of Cybermap Systems.

them to join his startup. Because these friends are not yet connected among themselves, the founder acts as a connector and a hub of trust. Picture 2 is the fully operational startup, with core team members fully connected with each other and exchanging multiple messages each day. The founder maintains a central role as the hub of trust, but each core team member communicates directly with all other core team members. Picture 3 illustrates the "dissolve" phase, in this case the beginning of the startup's demise. Communication has stalled; the founder is distracted from running the startup and has withdrawn from daily communication. He is now located at the team's periphery. While he is still fully connected, the rest of the core team is exchanging more messages without including the founder.

In this case, TeCFlow was used retrospectively to shed light on the inner workings of a startup company. The temporal plot of GBC and the movie of the communication flow convey deep insights into the rise and demise of this COIN that would otherwise have been difficult to obtain.

Project Management Pattern: Communication
in a Software Development Team

Our next example analyzes a Swiss e-banking project (discussed in chapter 5), one of the projects delivered by the Deloitte e.Xpert practice. The core team on Deloitte Consulting's side included some 20 consultants, led by a project manager and a project partner; the client team comprised the client project manager, a senior manager, and six subject matter experts. The analysis described here focuses on the phase of the project just before "go live" and the handover of software to the client. This period featured intense negotiations between client project management and senior managers of the consulting practice. There was also a change in Deloitte's project management.

Figures B.17 through B.20 are snapshots that illustrate the knowledge flow within the team during critical phases of this project period. Figure B.17 shows the communication flow as an addition to the original contract was negotiated, while the technical core team was hard at work implementing the technical system. On the figure's left side, Deloitte's legal team forms a dense cluster, intensively discussing contractual details. Leaders of both the consulting team and the client project team are in the center of the structure, communicating with everyone. The technical core team members

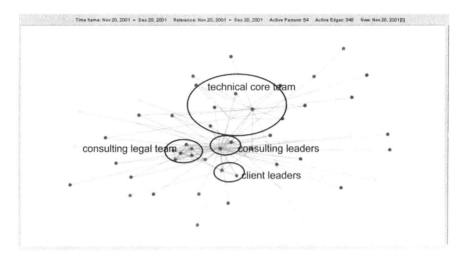

Figure B.17.   Contract negotiations with the client.

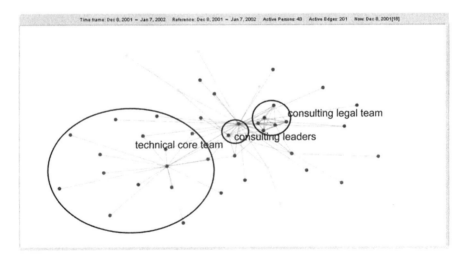

Figure B.18.   Testing of critical programming code.

are forming another COIN at the top of the figure, chiefly communicating among themselves.

During the subsequent testing phase shown in figure B.18, the test coordinator has a centralized role coordinating the developers. The legal team is still tightly clustered, while the clients (lighter dots) are (too) peripheral. In figure B.19, legacy data are converted from the old to the new system. The database administrator is clearly recognizable in the center of the technical core team. The legal team is no longer active, and client and consulting leaders collaborate somewhat more intensively. Finally, in the handover phase (figure B.20), client and consulting leaders collaborate closely, while the technical

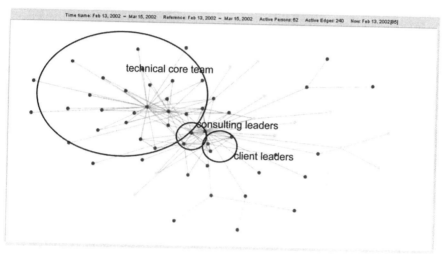

Figure B.19. Conversion of legacy data.

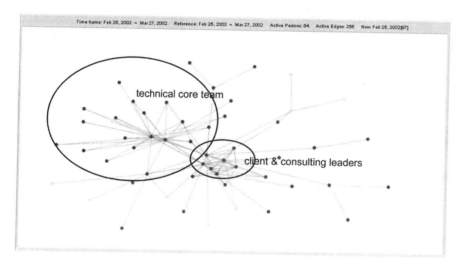

Figure B.20. Handover of entire application to client.

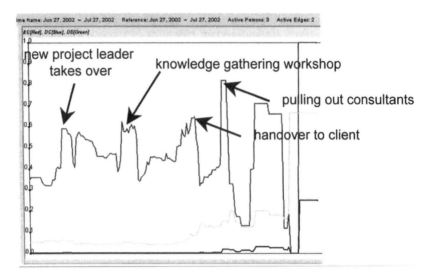

Figure B.21.   GBC of project team over 150 days.

core team forms a separate cluster. Collaboration between the client technical team and the consulting technical team is low.

Figure B.21 illustrates the main activities during these project phases. The project has generally "highly democratic" and low GBC. Major events become peaks in the curve. Project management takeover by the new leader can be identified clearly, as his initially more centralized communication style leads to a spike in GBC. The next peak in the curve comes from a knowledge-gathering workshop actively coordinated jointly by the project leader and the knowledge manager. Organizing handover as well as the final pullout of the consultants led to the next peaks.

Figure B.22 illustrates the communication patterns of the different actors in the project. We can see that the consulting project leaders receive more messages than they send, and that they are also the most active senders. The programmers receive more messages than they send, while the administrative staff members send more messages that they receive. This is fairly typical behavior for a commercial software project.

The findings from the TeCFlow analysis provide a fast and convenient way to pinpoint weaknesses in the project management process and can assist project managers to improve efficiency and communication flow. In this particular case, we can see several areas where TeCFlow could have helped.

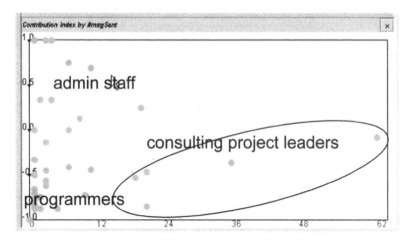

Figure B.22.   Contribution index of different people in the e-banking project.

- In the early phases of the project, senior leadership and technical staff on the consulting side were not well integrated. During the project, we had a major communication failure within the consulting team that resulted in a hardware crash that also corrupted the backup system. This led to an eight-week delay and an $800,000 cost overrun. Had TeCFlow been available, this communication breakdown would have been recognized earlier and might have been corrected before such major damage to the project.
- Initially, junior project team members who assumed critical roles were not appropriately recognized and rewarded. Also, potential staffing bottlenecks could not be identified and alleviated before serious problems arose. We had two instances in which the critical contributions of junior project members were not recognized in time to prevent problems; a major interpersonal crisis was avoided only by the last-minute intervention of senior management, which meant a multiweek delay. With TeCFlow, the contributions of the junior members would have been more obvious and bottlenecks would have been spotted, which could have saved two to four weeks of project time.
- The integration of senior team members who were not particularly strong communicators was not optimal. Additionally, there were some senior members who made no real contribution to the project.

This became obvious only after the TeCFlow analysis. By making more efficient use of technical staff during the project's lifetime, the consulting head count could have been reduced by three to five people.

- The client was only peripherally involved in the handover phase, and the client staff was reluctant to assume responsibility for the system, which delayed the project. TeCFlow would have pointed out this communication problem in real time and saved perhaps three to four weeks.

In this $10 million project, the transparency and knowledge flow optimization made possible by TeCFlow might have resulted in overall savings of at least 20 percent.

Sales Force Pattern: Large Account Management in a Consulting Firm

A final example illustrates the use of TeCFlow to improve the efficiency of a sales team. The data set consists of the e-mail archives of the new senior sales manager of a Deloitte e.Xpert account management team that served a single Fortune 500 client. The new sales manager was replacing a retiree. As his final assignment, the retiring senior sales manager introduced his replacement to the client.

The analysis is documented here as a series of snapshots of the TeCFlow movie automatically generated from the mailbox of the new senior sales manager. Figure B.23 illustrates the old sales manager introducing the new sales manager to Deloitte's account management team. In figure B.24, the old sales manager introduces his replacement to client executives. Consultants are represented by light dots; clients are represented by dark dots. The old sales manager is still more and better connected to the client executives than is the new sales manager, although new ties between the new sales manager and client executives are emerging.

Finally, figure B.25 illustrates coordination of a large proposal for the client. The new sales manager is at the center of the proposal team, while the old sales manager—who stuck around the company for another six months before fully retiring—is in a more peripheral role. At the same time, a group of consultants and client executives (the cluster of black dots in the upper

old sales manager

Time window: Nov 30, 2001 ~ Jan 18, 2002   Active Actors: 10   Active Edges: 17   Now: Nov 30, 2001[0]

0 .   58   116   174   232   290

new sales manager

Figure B.23.   Initial internal handover from old to new manager.

right) try to arrange a social event. The new sales manager is too preoccupied with proposal preparation and misses this opportunity to connect with a new group of potential customers.

Figure B.26 displays a history of changes in GBC and density, identifying the major sales opportunities of the Deloitte sales team. Troughs in the GBC curve indicate gatherings of a proposal team. As it happens, the new sales manager only found out about some of these proposals through the TeCFlow analysis; drilling down on the messages in the TeCFlow clusters showed that people had been working on proposals at these gatherings.

The contribution index plot (figure B.27) shows that the most committed Deloitte team member is a midlevel consultant working as a project manager at a client site. The figure also illustrates the somewhat passive communication behavior of the new sales manager; the old sales manager, although officially retiring from his function, is still communicating more actively. Figure B.27 also points out the two most active clients.

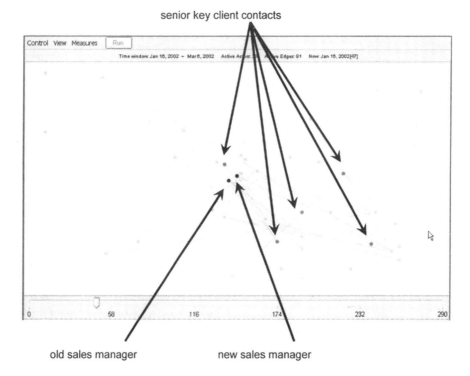

Figure B.24.   Introduction of new sales manager to senior client executives.

Had TeCFlow been available to the new sales manager, he could have systematically identified key clients and key proposal opportunities, thus substantially increasing the number of new opportunities for submitting proposals. The analysis uncovered several specific areas where TeCFlow could have helped.

- The old sales manager still held key client relationships even after the introduction of the new sales manager. The new sales manager could have been much more active in leveraging the connections of the retiring manager in developing his relationships.
- The new sales manager missed a unique opportunity to connect with a group of new client executives when they were trying to organize a social event.

new sales manager

old sales manager

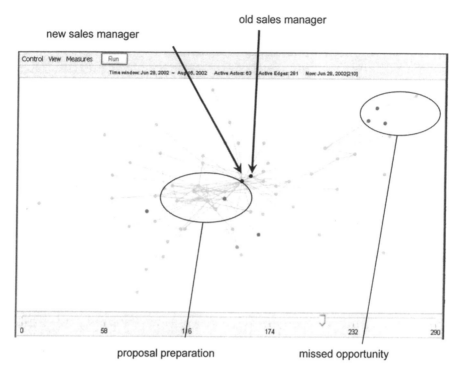

proposal preparation   missed opportunity

Figure B.25.   Preparation of a large proposal led by new sales manager.

- Coordination between different Deloitte sales teams was not optimal. The new sales manager missed multiple opportunities to support submission of proposals to the client.
- A midlevel consultant owned the best connections to the client. The sales manager could have made better use of these connections as a means of building new relationships.
- The most active customers were not known and, therefore, could not adequately be taken care of by Deloitte.

Summary

The five examples in this appendix show that TeCFlow analysis of the flow of knowledge offers a fast, convenient, and valuable way to discover different

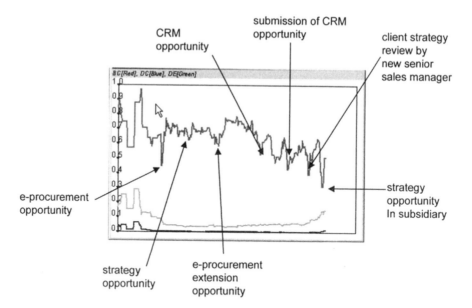

CRM
opportunity

submission of CRM
opportunity

client strategy
review by
new senior
sales manager

e-procurement
opportunity

strategy
opportunity
In subsidiary

strategy
opportunity

e-procurement
extension
opportunity

Figure B.26.   GBC of account management team.

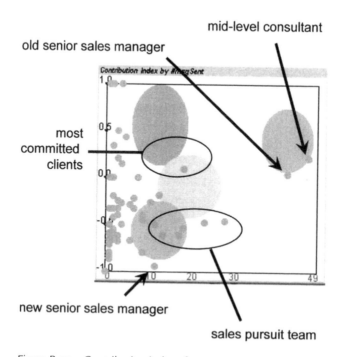

mid-level consultant

old senior sales manager

most
committed
clients

new senior sales manager

sales pursuit team

Figure B.27.   Contribution index of account management team.

phases in the life cycle of an online community. TeCFlow conveys insights that would be much more expensive to obtain by other means, and the TeCFlow-based approach makes it possible to find periods of low and high "democracy" (i.e., GBC) and to identify periods of high productivity and information dissemination. While this tool needs to be complemented by other contextual cues (such as interviews with community members and a content analysis of the messages exchanged) to obtain a full understanding of the activities, it is clearly a software tool with powerful applications for uncovering COINs and putting them to beneficial use in their hosting organizations.

Appendix C introduces knowledge flow optimization as a method for converting organizations into COINs. Knowledge flow optimization utilizes TeCFlow to find emerging clusters of people working together as innovation teams. It assists in creating a culture for COINs to convert team members into creators, collaborators, and communicators, thereby greatly improving organizational creativity, quality, and efficiency.

Knowledge Flow Optimization (KFO)

One summer morning in New Hampshire, I was hiking up Mount Lafayette from the Appalachian Mountain Club's Greenleaf Hut. The weather forecast had predicted nice, sunny weather. Suddenly, though, the sky became covered with dark clouds, and a few drops of rain began to fall. Because I had trusted the weather forecast, I was wearing only light clothes and had no heavy raincoat. Within five minutes, I was soaking wet.

As I got wetter and wetter, I began to reflect on the reliability of the weather forecast. Today's weather forecasting allows for predictions about weather phenomena based on observations of factors such as atmospheric pressure, wind speed and direction, precipitation, cloud cover, temperature, and humidity. In a weather forecasting system, weather patterns consisting of time series of recently collected data are fed into a computer model to predict the weather patterns of the next 5 to 15 days.

The parallels between weather forecasting and knowledge flow optimization (KFO) are striking. In KFO, communication patterns consisting of time series of collected communication data provide insights into complex group dynamics and make it possible to predict future group behavior (figure C.1). As with weather patterns used to predict sunshine and thunderstorms, communication flows allow for predicting positive and negative developments in groups of people. This requires a delicate combination of mathematical modeling, experience, and intuition. KFO is extremely valuable as an early

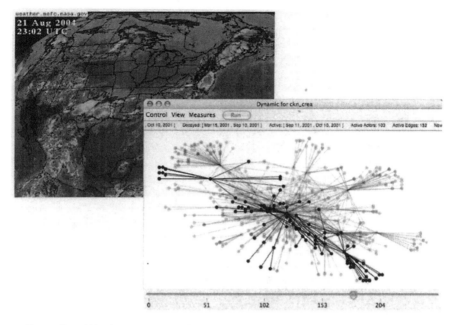

Figure C.I. Weather patterns and communication patterns allow for predicting the future.

warning system, showing high-pressure systems, impending storms, and other relationships in groups that are difficult to anticipate through other means. It offers insights into organizational dynamics. By analyzing and aligning business processes and knowledge flow, organizations get a unique opportunity to increase the productivity of knowledge workers through greater creativity, efficiency, and quality.

Business process reengineering forever changed the way companies do business, introducing a process focus and streamlining structured business processes. KFO does the same for unstructured, knowledge-intensive innovation processes. By visualizing the flow of knowledge, making it transparent, and reengineering its flow, organizations and individuals become more creative, innovative, and responsive to change. KFO offers companies a chance to complement their business process maps and organizational charts with much more fluid maps of relationships (figure C.2). By making the communication flow transparent, KFO can make existing business processes more efficient, allowing organizations to make better use of people by unburying them from conventional multilayer hierarchies. By establishing flexible ad hoc work

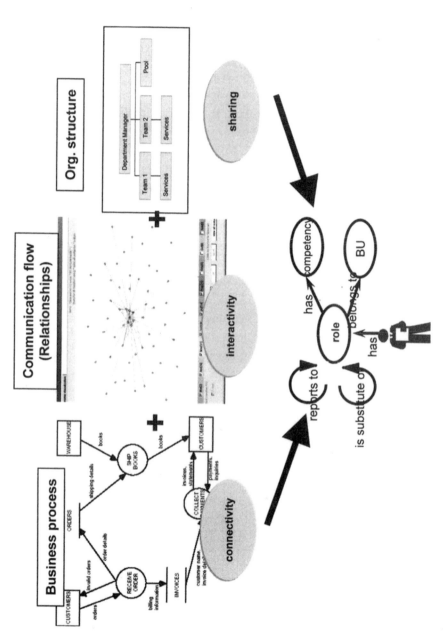

Figure C.2. KFO aligns business processes and organizational structure through communication flow analysis.

flows based on communication and relationships, people can become more efficient in their roles, which in turn will lead to greater individual motivation. Knowledge flow links people across organizational and process boundaries, and analysis of this knowledge flow gives an early indication of hidden innovation. Nascent creative communities can be uncovered by visualizing intra- and interorganizational communication patterns; business processes and organizational structures can then be aligned with the actual communication flow. KFO introduces a framework for the temporal analysis and visualization of communication patterns using the temporal communication flow analysis (TeCFlow) software tool introduced in appendix B. TeCFlow makes it possible for organizations to uncover traces of emerging collaborative innovation networks (COINs) by visualizing their communication flows.

There are four steps to KFO (figure C.3). First, communities are analyzed by TeCFlow and by interviewing organization members. Second, existing communication patterns are assessed on the degree to which they resemble the collaborative knowledge network (CKN) framework (i.e., innovation diffusion through the double helix communication pattern) and the three dimensions—interact, connect, share—of online behavior discussed in chapter 6 and appendix A. Third, the organization is optimized by applying the principles of COINs explained in earlier chapters. Fourth, communication flow is monitored continuously using TeCFlow.

KFO Step 1: Visualize Virtual Communities to Find COINs

The first step in KFO is to visualize the communication flow and make it transparent. The TeCFlow visualizer introduced in appendix B offers a convenient way to uncover communication flows using e-mail and phone logs, Web access archives, and chat session transcripts.[1] As figure C.4 shows, the TeCFlow tool addresses both the macro and the micro levels of teams. On the group or macro level, interactive movies and the plot of changes of group betweenness centrality (GBC) allow observers easily and quickly to distinguish different communication patterns over the lifetime of an online community. On the individual or micro level, interactive movies and the contribution index make it possible to extract individual communication patterns.

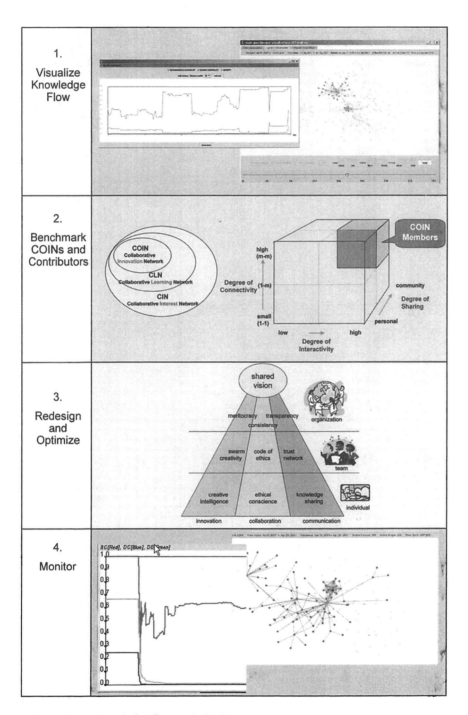

Figure C.3.   Knowledge flow optimization.

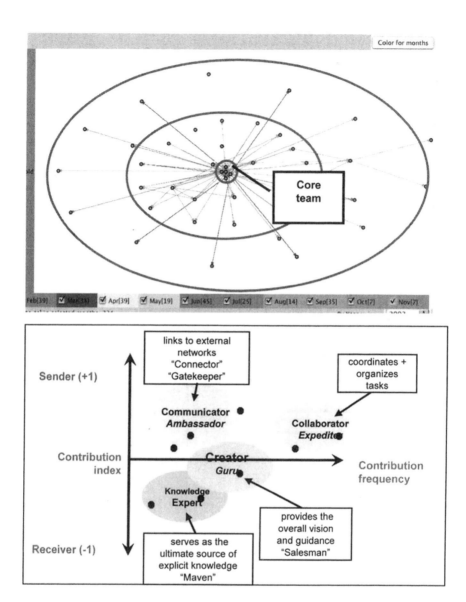

Figure C.4.   Macro- and micro-level communication pattern analysis. (Top panel) Macro level—communication patterns of groups . (Bottom panel) Micro level—communication patterns of individuals.

By generating interactive communication movies, we can identify typical communication patterns of COINs, collaborative learning networks (CLNs), and collaborative interest networks (CINs). Each of these types of communities has a distinctive structure; COINs have the strongest small-world, scale-free structure. The three TeCFlow snapshots in figure C.5 illustrate how a

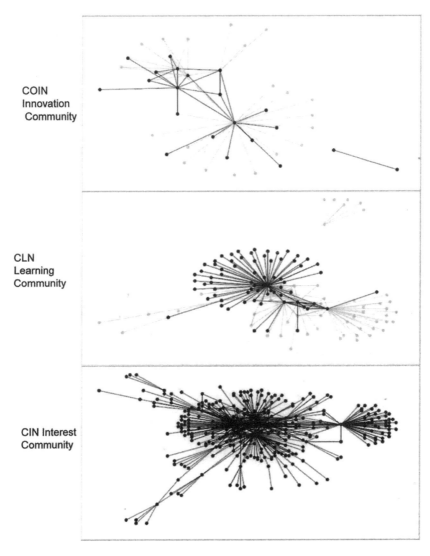

Figure C.5.   Snapshots of TeCFlow communication flow movie to identify COINs/CLNs/CINs.

COIN, a CLN, and a CIN grow over time and become identifiable (here, we use how the "CKN concept" was developed and spread out in the Deloitte e.xpert practice as our example). The top picture shows the COIN core team becoming clearly recognizable as a cluster in the center of the community. The middle picture shows the COIN growing into a CLN as more members join the periphery. The bottom picture shows the last TeCFlow snapshot of the same movie, illustrating the CIN structure of the entire community. The Deloitte e.xpert CIN includes the COIN and CLN, but also has many other members in the e.xpert practice who have not yet been infected by the "CKN virus."

The three-step TeCFlow analysis described in appendix B can also be used to analyze the roles of individuals in virtual communities; to help identify the experts within the organization; and to identify who has the most decision-making influence and who the innovators, collaborators, and communicators are. Calculating and plotting communication frequency and contribution index with TeCFlow is a quick and convenient way to find central COIN members.

In addition to mining communication archives, input for TeCFlow can be obtained through periodic online questionnaires administered to members of virtual communities or even to outsiders.[2] Figure C.6 is a sample online form on which community members can indicate the frequency of their e-mail, phone, and face-to-face communication.[3]

KFO Step 2: Benchmark COINs and Core Contributors

Once a COIN and its core contributors have been identified, performance should be measured to find areas for improving knowledge worker productivity—which can be a real challenge. There are three primary ways to assess a COIN's performance:

1. *Evaluate the COIN's output in financial terms.* Once a COIN has developed a tangible product, the value of that product can be measured. For the Deloitte e.xpert virtual consulting practice, we could measure its cost and revenue against operating a conventional consulting practice. For DaimlerChrysler's e³ e-procurement project (see chapter 5), realized savings could be compared to project costs.

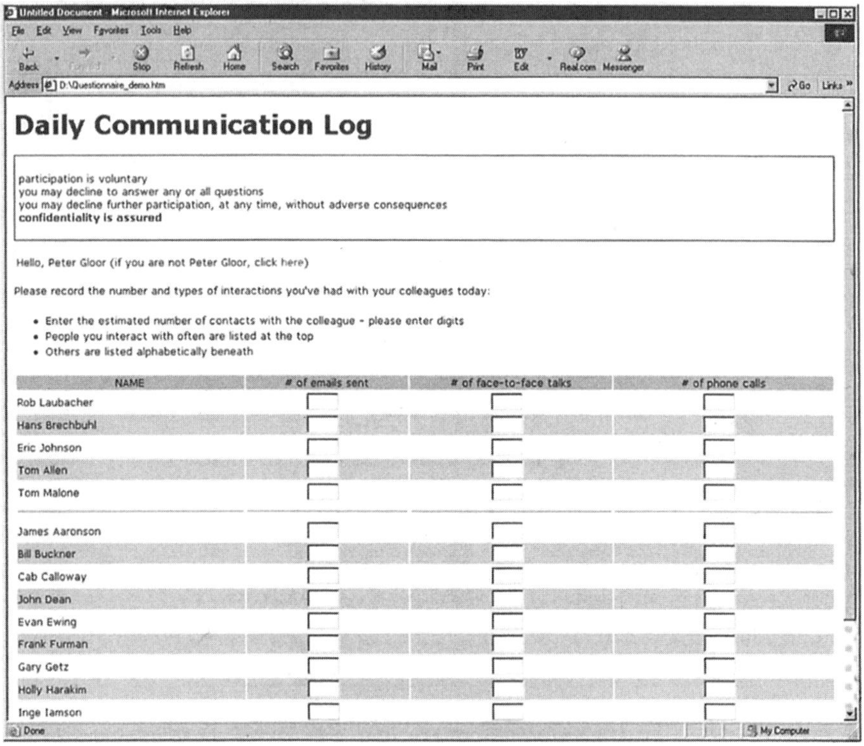

Figure C.6.   Web survey form.

The drawback of this method is that it can be difficult to correlate financial results with the efforts of a COIN.[4]

2. *Ask external experts to rate the COIN's performance.* Collecting the subjective evaluations of a large number of outsiders is another way to assess the impact of COINs. This can be done through online questionnaires, with the obvious drawback that such a method can absorb a lot of time and resources.

3. *Assess objective measures.* Using the CKN framework of innovation diffusion and the three dimensions of online behavior discussed in chapter 6 and appendix A, it is possible to measure the speed of innovation diffusion, how long it takes for the COIN's innovation to be accepted, and the level of that acceptance.[5] The basic idea is to correlate objective measures of COIN success such as the value

generated by the COIN with social network structure and individual communication behavior.

High-performing virtual teams exhibit distinctive networking properties. Social networking analysis metrics can be computed both for groups and individuals.[6] On the group level, these metrics identify typical patterns of COINs, CLNs, and CINs. Productive COINs use the CKN ecosystem to collaborate and to disseminate information. By exploiting correlations between the structure of the social network and metrics such as GBC and density, potentially high-performing group structures such as COINs can be found. An effective COIN will exhibit a combination of high density with low GBC over sustained periods of time.

By measuring the position of individuals in networks, we can draw conclusions about their performance. Using the contribution index function of TeCFlow, the three dimensions of online behavior—again, interactivity, connectivity, and sharing—can be measured for each team member; this will determine the individual's role as creator, collaborator, communicator, or knowledge expert.

KFO Step 3: Redesign and Optimize

COIN creation is a fluid process—a COIN cannot be mandated into action. The progression of COIN growth is similar to chemical crystallization, where a nurturing liquid and a crystallization germ are enough to start the process. Just as a crystal grows from the germ on its own by adding more molecules through chemical attraction forces, the trigger for COIN members to join the community stems from their own motivation. Growing the largest crystals with the most desirable properties requires a thorough understanding of the procedure and ingredients of the crystallization process. The same principle applies to the COIN growth process (although in the case of COINs, it may be better to have multiple, smaller COINs).

Seven success factors are involved in bringing a COIN to fruition. It is not necessary to accomplish them in sequence, but each one is critical.

1. Establish swarm creativity and give up central control.
2. Nurture the critical roles of creator, communicator, and collaborator.

3. Establish distributed trust.
4. Establish a common code of ethics.
5. Establish a small-world structure of high connectivity, interactivity, and knowledge-sharing.
6. Set up a collaborative Web workplace.
7. Know when to change the organizational structure.

Senior executives in organizations that host COINs *must* create an environment in which their COINs have the opportunity to achieve all seven of these critical success factors, and in which all COIN members can become the best at playing whichever COIN role they fill. The potential benefit is enormous: imagine getting an intrinsically motivated group of people to deal with a tough problem not for promotion or a pay rise, but out of fascination

---

Sidebar C.1
Making COINs Happen

Throughout this book, our focus has been on the existence of COINs (even if they are hidden from view) and on the conviction business organizations would be wise to foster a culture supportive of COINs in order to reap their benefits. Most cases presented in chapter 5 describe COINs that seem to have emerged in the course of work—that is, that seem to have happened "naturally."

This section departs from that focus to take a bold leap and propose that there are actual *steps* that can be taken to *make COINs happen*.

The ideas behind how to create COINs actively are not yet fully worked out. Companies are trying out the COIN concepts and KFO, and their experiences are forming the basis for further analysis. In my own project work, I am testing whether it is possible to *generate* swarm creativity (not simply create the best conditions for swarm creativity) by taking great pains to ensure that only intrinsically self-motivated people who share the same "DNA" (with respect to being potential COIN members) comprise a team. This also involves actively ensuring that each location has a hub of trust, and making sure, as well, that the hubs of trust get to know each other and establish mutual trust in a face-to-face meeting (or meetings).

Each of the seven critical success factors in KFO step 3 involves *action,* and therein lies the bold leap. Can COINs *actively be created?* The evidence thus far strongly suggests "yes."

with the problem and a shared vision for coming up with an innovative solution.[7]

Critical Success Factor 1: Establish Swarm Creativity and Give Up Central Control

Only if senior executives are willing to give up centralized control and ready to trust the self-organizing capabilities of COINs will they harvest the full benefits of COINs. Recruiting the brightest and best, and then letting them loose on their own, requires high levels of confidence and managerial constraint. The same goes for investors, who expect centralized control over processes to "ensure" a return. But if investors want to leverage innovation by COINs, they will have to take a leap of faith and embrace swarm creativity and self-organization.[8]

Critical Success Factor 2: Nurture Creators, Collaborators, and Communicators

Growing a CKN ecosystem requires skills different than those managers typically learn in business schools. Managers need to emphasize putting together well-functioning teams with members who have the right "genes" to "swarm" around their leaders.

A team of creators, collaborators, and communicators does not need much centralized control. Rather, it needs strategic guidance and an environment that offers the cultural and technological prerequisites (mutual trust and necessary Internet collaboration tools at the disposal of every team member) for a CKN ecosystem. Once core team members have been identified, the team should be managed as an X-team (see chapter 5), with members changing their roles according to project needs. Accepting a continuously changing team, in which knowledge experts become creators, collaborators, and communicators, and in which core team members withdraw to become external experts, allows a flexible online community to grow, continually building high mutual trust and a distinct online identity.

Managers who insist on substituting centralized control for trust will drive COINs underground. There, they may continue to grow, but the chances are greatly reduced that the hosting organization will be able to reap the benefits.

Critical Success Factor 3: Establish Distributed Trust

For COINs to succeed, it is essential to create and maintain trust among COIN members. This is achieved most efficiently by meeting face-to-face, but can happen remotely—although it will take more time. If a group of people works together over long distances and an extended period, and if team members consistently deliver high-quality work, trust will develop remotely.

Sharing project-relevant knowledge and maintaining a transparent work environment are important elements of building and preserving a high level of trust, as is developing a shared vision and common goals. When team members participate in shaping the vision and goals of the community, the result is substantial buy-in[9] (which builds trust). The process of adjusting and fine-tuning goals should also be conducted in public, with an opportunity for all COIN members to collaborate so that the entire community identifies with the goals.

Critical Success Factor 4: Establish a Common Code of Ethics

When people are fully committed and intrinsically motivated, they will implicitly abide by a common ethical code when they work together. They will apply the Tao of COINs (see chapter 4) without having to learn it. Leaders who demonstrate respect for the individual and who obey the Golden Rule will serve as role models for the entire community.

Nevertheless, making the Tao of COINs explicit, and customizing it to the local environment, will clarify the rules for new COIN members and accelerate the process by which they learn the "right behavior" from COIN "elders."

Critical Success Factor 5: Establish a Small-World Structure
of High Connectivity, Interactivity, and Knowledge Sharing

A CKN ecosystem interoperates as a scale-free, small-world network in which teams are mutually connected by leaders acting as gateways and hubs of trust. Having strong leaders act as local hubs of trust will build an efficient, flexible, and robust network of global scalability. Local leaders should know and trust each other. Under ideal circumstances, leaders at different locations locally recruit new core team members whom they already know and trust.

The fastest way to build trust is to meet other core team members face-to-face. If this is too expensive, a collaborative Web workspace provides the technical means for COIN members to work together over the Internet and to develop confidence in the skills of other COIN members through reliable cooperation over time. At present, the e-mail component of the collaborative Web workspace is the primary means of communication for a geographically distributed COIN.

Critical Success Factor 6: Set Up a Collaborative Web Workspace

By combining simple Web-based collaboration tools such as e-mail, blogs, wikis, and Web conferencing, an organization can build a collaborative Web workspace to support COINs. COINs should establish a persistent organizational memory on the Web that is easily accessible to all members to share documents.[10] Weblogs, wikis, and mailing lists offer a cheap and efficient way to establish that group memory.[11]

A protocol for successful e-mail communication in COINs has a few simple rules: leaders of the community must be easily accessible by e-mail, and COIN members should respond quickly to messages (or indicate their temporary unavailability if they are unable to respond in a timely manner).

Videoconferencing, which approximates the intimacy of face-to-face meetings at lower cost, has become increasingly common at large companies. Smaller companies and individuals can also rent videoconferencing capabilities. But while useful for a meeting now and then if people cannot be brought together in one location, frequent use of videoconferencing is no substitute for building trust through face-to-face meetings, and hence is not worth the price. Web conferencing, though, is a cost-efficient substitute for bringing people together who have already worked together in the past and want to meet regularly. By using document-sharing software such as Placeware, meeting participants can share one screen virtually, while public and private chat windows allow side communications. Low-cost Webcams, or even a conference call, can be used to discuss what is shown on the shared screen. For addressing a large group, Webcasts can be useful—for instance, if

Sidebar C.2
An Ideal Wiki Application

Wikipedia is the best-known wiki application, allowing people with very few computer skills to collaborate and share knowledge on the Web. The figure presented here illustrates how easy it is to change an entry in Wikipedia. The top panel shows the Wikipedia entry for Tim Berners-Lee as seen by the reader. By clicking on the "edit this page" tab, a user can open the entry for editing (the middle panel). Clicking on the "discussion" tab (bottom panel) brings up the page on which authors can sort out disagreements about the text on the main entry page.

Wikipedia entry for Tim Berners-Lee.

a CEO wants to address all of a company's employees, or to transmit keynote speeches of a conference.

Critical Success Factor 7: Know When to Change to a Conventional Organizational Structure

The final key to success is to recognize when it is time for a change. Once a COIN has developed a commercial product, the organizational form of the COIN should be changed into a project team or a business unit. At the same time, the CKN ecosystem developing around the COIN should be leveraged according to the double helix of innovation. Disseminating innovations from COIN to CLN to CIN is a cost-efficient way to gain a large audience for the work products of the COIN. The learning network of the CLN is instrumental in recruiting new core team members. The interest network is the perfect mechanism for disseminating the innovation. Lurkers at the periphery of the CIN are the agents of knowledge dissemination, spreading the seeds of acceptance for the new product. At the same time, the CIN is the incubator of new innovation, creating an environment in which new COINs can crystallize and grow.

KFO Step 4: Monitor COINs

Maintaining a culture for COINs is not a one-time activity. Once the knowledge flow of the organization has been optimized according to the seven critical success factors above, the ingredients for a COINs-supportive culture should be constantly supervised, gauged, adjusted, and remixed if necessary. Hence, step 4 of KFO is the continuous monitoring and nurturing of a culture for COINs.

As a communications cockpit, TeCFlow provides an excellent mechanism for monitoring communication flow based on e-mail, chat, phone, and Web access traffic. In this way, an organization can produce "weather forecasts" that reveal "high-pressure systems" and impending "storms." The communication patterns of such forecasts identify emerging COINs, pinpointing clusters of high communication traffic as indicators of new innovation, as well as of groups that need closer attention.

Realignment of business processes and knowledge flow offers a great chance to individuals to become better creators, collaborators and communicators. But there are also steps that individuals can take themselves to become more effective COIN members.

How to Be a Successful COIN Member

Linux is an ecosystem, and in this ecosystem there is fast growth vegetation and slow growth vegetation. The fast growth vegetation are the people who took what had already been done by Unix, and without changing its design they copied it while making coding improvements.

Then there are those who look at Linux, and see something just barely begun that needs a complete overhaul. These are the slow growth vegetation.[12]

External fame and reputation, or recommendation by a trusted community leader, can jump-start the initial position and career of a new COIN member. A trusted position *within* a COIN, though, must be earned over time. Senior-level COIN membership comes only to those who deliver high-quality work over an extended period. Joining a COIN because of an intrinsic motivation is the road to sustainable success, and tackling hard problems—"slow-growth vegetation"—is the best way to gain an online reputation in the community. As Hans Reiser puts it, exemplary COIN members are "people who chose trying to create a better society as their life's work at a substantial cost in personal income."[13]

A good online reputation acquired through hard work is easy to lose. By acting sloppy, being unreliable, or breaking the ethical code, a COIN member will quickly diminish his or her reputation. The way to become a successful COIN member and *sustain* that success is to follow the rules of swarm creativity, ethical behavior, and distributed trust.

KFO Applications

Knowledge-intensive organizations that optimize the flow of knowledge realize strategic advantages. Consulting firms, software development projects, project management of large projects, mergers and acquisitions, and sales forces are among the application areas where KFO is ideally suited to improve organizational creativity, quality, and effectiveness.

Get the most strategic value from mergers and acquisitions. The Daimler-Chrysler e^3 initiative (see chapter 5) is a prime example of how an organization can derive tremendous strategic and financial value by redesigning its knowledge flow. While the initiative began as a project to reengineer the business processes of the merged Daimler and Chrysler procurement departments, it quickly became much more. The initiative turned the entire business model of procurement upside down and created a multitier online marketplace for suppliers of auto parts, unleashing tremendous value for the enterprise. The e^3 initiative team at DaimlerChrysler operated as a genuine COIN, with members from both sides of the Atlantic adhering to the Tao of COINs, creating the new solution in true swarm creativity, and collaborating in a highly efficient small-world networking structure with the senior project members acting as hubs of trust. The DaimlerChrysler case is an exemplary illustration of the strategic advantages organizations in merger situations can expect by the internal transparency, consistency, and meritocracy of COINs.

Optimize research and development. Research and development organizations that redesign their knowledge flow so that it operates as a network of COINs embedded in a CKN ecosystem can expect substantial advantages. The development of the CKN service offering for Deloitte Consulting (see chapter 5) illustrates how an innovative new product was developed by a COIN that was recruiting new members from its surrounding CLN while using its CIN as a sounding board and sales and marketing mechanism.

Streamline project management. The e-banking project analysis in appendix B shows how applying TeCFlow to redesigning the flow of knowledge would have improved the efficiency of a multimillion-dollar project by at least 20 percent. Monitoring project management communication for better quality of project output translates into substantial savings. The KFO approach offers myriad communication-related benefits: it greatly reduces communication failures among project members; it converts one-way communication into two-way dialogues; and it reveals core contributors as well as lurkers. Changing project culture to a COIN-based approach makes the teamwork more efficient, unlocking the creative potential of neglected team members. Visualizing the flow of knowledge makes it easier to find good ideas within the organization.

Improve the sales process. The sales force communication pattern analysis at the end of appendix B shows how applying KFO improves the efficiency and productivity of sales and marking. TeCFlow makes it possible to find who is most productive and unproductive on a sales and marketing staff.[14]

NOTES

Introduction

1. Gladwell, *The Tipping Point: How Little Things Can Make a Big Difference* (2002).
2. Now Sir Tim Berners-Lee, he was knighted in 2003 by Queen Elizabeth II of the United Kingdom for his contributions to the creation of the World Wide Web.
3. Bush, *As We May Think* (1945).

Chapter 1

1. Hamel, *Waking Up IBM: How a Gang of Unlikely Rebels Transformed Big Blue* (2000).

Chapter 2

1. Adapted from Bonabeau, Dorigo, and Theraulaz, *Swarm Intelligence: From Natural to Artificial Systems* (1999).
2. See Wikipedia at www.wikipedia.org, Encarta at encarta.msn.com, and *Encyclopaedia Britannica* at www.britannica.com.
3. Alex Halavais, a University of Buffalo communication professor, conducted an experiment to address concerns about the accuracy and scientific validity of an encyclopedia that relies entirely on volunteers for its updates. He entered mistakes into 13 Wikipedia entries to see if his "hacker" attack would be detected and to gauge how long it would take the volunteers to fix the mistakes. His optimistic expectation was that the errors would be fixed within 24 hours; however, he was surprised to find that it took the Wikipedia cyberteam only a few hours to correct all his mistakes. For more information, see alex.halavais.net/news/index.php?p=794. Halavais's findings have been confirmed

by IBM researchers who looked at controversial articles, such as the entry on "abortion." They found that while entries were sometimes purposefully altered or even deleted, it usually just took only minutes for the malevolent alterations to be fixed. For more information, see www.research.ibm.com/history/index.htm.

4. See Challet and Du, *Closed Source versus Open Source in a Model of Software Bug Dynamics* (2003); Raymond, *The Cathedral and the Bazaar* (1999); Stark, "The Organizational Model for Open Source" (2003); Paulson, Succi, and Eberlein, "An Empirical Study of Open-Source and Closed-Source Software Products" (2004).

5. Christensen, *The Innovator's Dilemma: The Revolutionary National Bestseller That Changed the Way We Do Business* (2000).

6. An argument can be made that the Web and Linux were nondisruptive, as there was already a system called "gopher" available which was very similar to the Web. UNIX was also available well before Linux. But by making the Web and Linux freely available, Tim Berners-Lee and Linus Torvalds were able to rally widespread support for their causes, which in turn had a disruptive effect on their respective industries.

7. Von Hippel, *The Sources of Innovation* (1997 [1988]).

8. Chesbrough, *Open Innovation: The New Imperative for Creating and Profiting from Technology* (2003).

9. Shneiderman, *Leonardo's Laptop* (2002).

10. Licklider, "Man-Computer Symbiosis" (1960).

11. Digital Systems Research Center, *The Tao of IETF: A Guide for New Attendees of the Internet Engineering Task Force* (1994).

Chapter 3

1. In his seminal work, Francis Fukuyama describes the basic lack of trust outside of family ties in Confucian Chinese society. See Fukuyama, *Trust: The Social Virtues and the Creation of Prosperity* (1995).

2. Contrast Marco Polo's writings with the writings of Sir John Mandeville. Making himself the focus of his book, Mandeville claimed to have been a mercenary for the Sultan of Egypt and to have served the Emperor of China in his war against a Southern Asian king. Descriptions of the Cyclops and people with the heads of dogs are some of the less outrageous of Mandeville's fantastic claims. See Mandeville, *Travels of Sir John Mandeville* (1984 [1356]).

3. Hancock, Thom-Santelli, and Ritchie, "Deception and Design: The Impact of Communication Technologies on Lying Behavior" (2004).

4. Joinson, *Understanding the Psychology of Internet Behaviour: Virtual Worlds, Real Lives* (2003).

5. Gates, *Business @ Speed of Thought, Using a Digital Nervous System* (1999).

6. It should be noted that leaders of Microsoft's competition, such as Sun Microsystems CEO Scott McNealy, are very critical of this approach to innovation and accuse

Microsoft of outright intellectual property robbery. When Microsoft created the same "look and feel" as the Macintosh user interface through 189 identical elements in its Windows operating system, Apple took the company to court—and lost.

7. Saxenian, *Regional Advantage: Culture and Competition in Silicon Valley and Route 128* (2000). Apple has learned this lesson, while Microsoft is increasingly closing its own interfaces to other software vendors. For example, Microsoft is currently involved in a lawsuit in the European Union, having been sued by vendors of operating system add-ons such as digital audio and video players. Competitors, including Real-Audio, claim Microsoft is gaining an unfair advantage by tightly bundling its own Media Player software with its operating system, repeating a marketing gimmick that worked perfectly in the past to stop PC buyers from using competitor Netscape's Web browser. This shows Microsoft is becoming extremely selective in its approach to transparency and information sharing.

8. SHARE's success thus far is impressive. In its first four years, SHARE linked *hundreds* of entrepreneurs, startups, and researchers. In 2003, it organized more than 60 events and hosted more than 4,500 visitors. SHARE also makes heavy use of the Web to support collaboration among geographically dispersed innovative knowledge workers—for example, by offering live Web transmission of its networking events.

Chapter 4

1. Kanter, *Commitment and Community: Communes and Utopias in Sociological Perspective* (1972), p. 73.

2. Digital Systems Research Center, *The Tao of IETF: A Guide for New Attendees of the Internet Engineering Task Force* (1994).

3. Gillmor, "In the Wild West of the Internet, There Are Good Guys and Bad Guys" (2003).

4. Rawls, *A Theory of Justice* (1999), p. 86.

5. Watts, *Tao: The Watercourse Way* (1977).

6. Digital Systems Research Center, *The Tao of IETF*.

7. Moon and Sproull, "The Essence of Distributed Work: The Case of the Linux Kernel" (2000).

8. Putnam, *Bowling Alone: The Collapse and Revival of American Community* (2000).

9. See http://smallworld.columbia.edu/.

Chapter 5

1. In 2004, Deloitte Consulting merged with the consulting services of Deloitte & Touche, and the entire firm has been renamed Deloitte.

2. The breadth of this effort was very wide, encompassing consultants at offices in Amsterdam, Brussels, Copenhagen, Frankfurt, Helsinki, Lisbon, London, Luxembourg, Madrid, Milan, Munich, Oslo, Paris, Rome, Stockholm, Vienna, and Zurich.

3. Williams and Cockburn, "Agile Software Development: It's about Feedback and Change" (2003).

Chapter 6

1. For example, Web conferencing tools such as Placeware (purchased by Microsoft in 2003 and renamed LiveMeeting) allow the participants to see who the other listeners are.

2. Research has demonstrated that the quality of group performance of a distributed team is directly proportional to the degree of knowledge-sharing among group members. See Cummings and Cross, "Structural Properties of Work Groups and Their Consequences for Performance" (2003).

Appendixes

1. Licklider and Taylor, "The Computer as a Communication Device" (1968).

Appendix A

1. These communities are sometimes called "practice communities" or "knowledge stewarding communities." See Wenger, McDermott, and Snyder, *Cultivating Communities of Practice* (2002).

2. Gardner and Gardner, "Your Sneak Peek into Motley Fool Stock Advisor" (2002).

3. CKN is the name of a Deloitte Consulting service offering based on an earlier and much less fully developed version of the collaborative knowledge network concept described in this appendix. At the time Deloitte offered this service, for instance, there was no distinction between COINs, CINs, CLNs, and CKNs. How Deloitte's CKN service was created is actually a great example of how a COIN works, as described later in this section.

4. See www.deloitte.com/dtt/cda/doc/content/collaborativeknowledge.pdf.

5. Deloitte's CKN initiative won considerable recognition from industry analysts and in the press. Articles about this work appeared in German, Austrian, Finnish, English, and United States newspapers. More important, the tools, concepts, and methodologies developed by the CKN initiative opened doors to potential clients across Europe, the United States, and Southeast Asia. As I was leaving Deloitte at the end of 2002, CKN-related projects were under way in Finland, Switzerland, Austria, the United States, the United Kingdom, Singapore, and elsewhere.

Appendix B

1. TeCFlow, a visual tool for temporal communication flow analysis, is currently being developed as part of a CKN joint project between the MIT Center for Coordination Science and the Center for Digital Strategies at Dartmouth's Amos Tuck School of Business Administration.

2. One could reasonably argue that measuring e-mail is insufficient for approximating social ties, and that this should be complemented by phone logs and transcripts of face-to-face interaction. In experiments by others, though, e-mail *has* been confirmed to be a generally reliable indicator and proportional to the other interactions.

3. In this particular working group, the W3C board appointed some leaders. As we shall see, when some of the appointed leaders failed to fulfill their leadership roles, other *non-appointed* people stepped in to lead the group—in typical COIN-like fashion.

Appendix C

1. To find the hidden COINs in organizations, a combination of mining communication archives with TeCFlow and interviewing executives and members of the organization is recommended.

2. The main advantage of online questionnaires is they make it relatively easy to collect and combine communication—*assuming users respond* (which is the potential drawback). By contrast, data such as e-mail, phone logs, and chat transcripts can be collected automatically without requiring that users enter data on their own.

3. Respecting individual privacy is of the utmost importance for the success of a COIN communication assessment. To obtain the buy-in of the study population, there must be assurance that the identity of individuals will be protected as data are collected, and that analysis results will be reported only in an anonymous format.

4. Admittedly, it can be very difficult to measure financially a COIN's output—but it is crucial. Some past efforts to do so have looked at productivity increases by measuring revenue generated in comparison to costs before and after KFO. This approach is very complicated, though, and so some simpler process-specific measures are advised: savings in reduced work hours, new revenue generated, reduced inventory costs, reduced production costs, etc.

5. Consider, for example, that fax machines took 100 years to be accepted and used by a sizable portion of the population; e-mail took 10 years; and the Web took five years.

6. Bulkley and Van Alstyne, "Why Information Should Influence Productivity" (2004). See also Cummings and Cross, "Structural Properties of Work Groups and Their Consequences for Performance" (2003).

7. The innovations—often new products—that COIN members create must meet two related criteria: (a) they must be virtually transportable from one work location to the other, and (b) to be virtually transportable, they must demand little local context.

8. COIN innovation has flourished for the past 20 years in the creation of new software, but today products as diverse as cars, airplanes, drugs, and movies are developed virtually on computer monitors. Unlike software, though, these products still require massive investments in production (e.g., factories, laboratories, studios). Hence the need for investors to embrace the COINs concept, too.

9. It is also important to secure the buy-in of external sponsors and the organization's top management as soon as possible. COIN members participate because they are intrinsically motivated, but they also depend on organizational acceptance. The best way to secure organizational acceptance is to recruit a fully committed sponsor high up in the organization. This is not a trivial task: the goals of a COIN are often, at first, of only peripheral interest to the hosting organization. The key to success is to persevere, perhaps even enlisting the support of external promoters who have the attention of top management of the hosting organization.

10. For example, the group-relevant part of a COIN's internal e-mail communication can easily be stored as a mailing list.

11. Weblogs, which have come to be known as *blogs,* are Web-based online diaries where anyone who can type can share thoughts. Blog readers can post comments on what they read. *Wikis* take this concept even further, allowing everyone to edit the contents of a Web page. Tools such as *twiki* (a user-friendly version of a wiki available for free at www.twiki.org/) allow for sharing all types of documents and for coordinating collaboration and workflow on documents through check-in and check-out. While word processing tools such as Microsoft Word allow users to compare and merge different versions of text documents, these simpler Web-based tools are entirely sufficient for coauthoring large documents.

12. Reiser, interview (2003).

13. Ibid.

14. Research has demonstrated that high-performing sales force members communicate more with external people than do average or low performers; that high performers make greater use of communication technologies for their work; and, surprisingly, that there is no correlation between performance and overall volume of communication. This means that very active communicators are not necessarily high performers. Additionally, there is a positive correlation between individual performance and short e-mail messages. By applying insights such as these to the communication patterns extracted by TeCFlow, the individual communication behavior of sales and marketing teams can be optimized for superior performance. See Bulkley and van Alstyne, "Does E-Mail Make White Collar Workers More Productive?" (2004).

REFERENCES

Ancona, D., Bresman, H., and Kaeufer, K. (2002). "The Comparative Advantage of X-Teams." *MIT Sloan Management Review* 43:3, 33–39.

Bonabeau, E., Dorigo, M., and Theraulaz, G. (1999). *Swarm Intelligence: From Natural to Artificial Systems.* Santa Fe Institute Studies in the Sciences of Complexity. New York: Oxford University Press.

Bulkley, N., and van Alstyne, M. (2004). "Does E-Mail Make White Collar Workers More Productive?" In *Proceedings of the North American Association for Computational Social and Organization Science* (NAACSOS), June 27–29, Pittsburgh, PA. Available online at www.casos.cs.cmu.edu/events/conferences/2004_proceedings/Bulkley_Nathaniel.pdf.

Bulkley, N., and van Alstyne, M. W. (2004). "Why Information Should Influence Productivity." In *The Network Society: A Cross-Cultural Perspective,* ed. M. Castells. Cheltenham, U.K.: Edward Elgar. Available online at http://ssrn.com/abstract=518242.

Bush, V. (1945). "As We May Think." *The Atlantic Monthly* 176:1, 101–108.

Challet, D., and Du, Y. L. (2003). "Closed Source versus Open Source in a Model of Software Bug Dynamics." Preprint. Available online at http://arxiv.org/abs/cond-mat/0306511.

Chesbrough, H. W. (2003). *Open Innovation: The New Imperative for Creating and Profiting from Technology.* Boston: Harvard Business School Press.

Christensen, C. (2000). *The Innovator's Dilemma: The Revolutionary National Bestseller That Changed the Way We Do Business.* New York: HarperCollins.

Cormen, T., Leiserson, C., and Rivest, R. (1990). *Introduction to Algorithms.* Cambridge, Mass.: The MIT Press.

Cummings, J., and Cross, R. (2003). "Structural Properties of Work Groups and Their Consequences for Performance." *Social Networks* 25:3, 197–210.

Digital Systems Research Center. (November 1994). *The Tao of IETF: A Guide for New Attendees of the Internet Engineering Task Force.* RFC 1718. Updated August 2001. Available online at www.ietf.org/tao.html.

Fukuyama, F. (1995). *Trust: The Social Virtues and the Creation of Prosperity.* New York: Free Press.

Gardner, D., and Gardner, T. (July 22, 2002). "Your Sneak Peek into Motley Fool Stock Advisor." *Motley Fool Stock Advisor* [online only]. Available online at http://investorplace.com/free/mfsa_free_002A.php.

Gates, W. (1999). *Business @ Speed of Thought, Using a Digital Nervous System.* New York: Warner Books.

Gillmor, D. (September 28, 2003). "In the Wild West of the Internet, There Are Good Guys and Bad Guys." *San Jose Mercury News.* Available online at www.siliconvalley.com/mld/siliconvalley/business/columnists/6881523.htm.

Gladwell, M. (2002). *The Tipping Point: How Little Things Can Make a Big Difference.* Boston: Back Bay Books.

Gloor, P., Dynes, S., and Lee, I. (1993). *Animated Algorithms.* CD-ROM. Cambridge, Mass.: The MIT Press.

Goleman, D. (1997). *Emotional Intelligence: Why It Can Matter More Than IQ.* New York: Bantam Books.

Hamel, G. (July/August 2000). "Waking Up IBM: How a Gang of Unlikely Rebels Transformed Big Blue." *Harvard Business Review,* 137–146.

Hancock, J. T., Thom-Santelli, J., and Ritchie, T. (2004). "Deception and Design: The Impact of Communication Technologies on Lying Behavior." In *Proceedings of the SIGCHI Conference on Human Factors in Computing Systems,* 129–134. New York: ACM Press.

Joinson, A. N. (2003). *Understanding the Psychology of Internet Behaviour: Virtual Worlds, Real Lives.* Hampshire, U.K.: Palgrave Macmillan.

Kanter, R. M. (1972). *Commitment and Community: Communes and Utopias in Sociological Perspective.* Cambridge, Mass.: Harvard University Press.

Leary, W. E. (April 14, 2004). "Better Communication Is NASA's Next Frontier." *New York Times,* A24.

Licklider, J. C. R. (March 1960). "Man-Computer Symbiosis." *IRE Transactions on Human Factors in Electronics* HFE-1, 4–11.

Licklider, J. C. R., and Taylor, R. (April 1968). "The Computer as a Communication Device." *Science and Technology,* pages unavailable.

Malone, T. W. (2004). *The Future of Work: How the New Order of Business Will Shape Your Organization, Your Management Style, and Your Life.* Boston: Harvard Business School Press.

Mandeville, J. (1984 [1356]). *Travels of Sir John Mandeville.* London: Penguin Books.

Moon, J. Y., and Sproull, L. (November 2000). "The Essence of Distributed Work: The

Case of the Linux Kernel." *First Monday* 5:11. Available online at www.firstmonday.dk/issues/issue5_11/moon/.

Paasivaara, M., Lassenius, C., and Pyysiäinen, P. (2003). *Communication Patterns and Practices in Software Development Networks.* Helsinki, Finland: Helsinki University of Technology.

Paulson, J., Succi, G., and Eberlein, A. (April 2004). "An Empirical Study of Open-Source and Closed-Source Software Products." *IEEE Transactions on Software Engineering* 30:4.

Putnam, R. (2000). *Bowling Alone: The Collapse and Revival of American Community.* New York: Simon & Schuster.

Rawls, J. (1999 [1971]). *A Theory of Justice.* Cambridge, Mass.: Harvard University Press.

Raymond, E. (1999). *The Cathedral and the Bazaar.* Sebastopol, Calif.: O'Reilly Media. Available online at www.catb.org/~esr/writings/cathedral-bazaar/.

Reiser, H. (June 18, 2003). Interview on Slashdot.org. Available online at http://interviews.slashdot.org/article.pl?sid=03/06/18/1516239.

Saxenian, A. (2000). *Regional Advantage: Culture and Competition in Silicon Valley and Route 128.* Cambridge, Mass.: Harvard University Press.

Shneiderman, B. (2002). *Leonardo's Laptop.* Cambridge, Mass.: The MIT Press.

Stark, M. (July 7, 2003). "The Organizational Model for Open Source." *HBS Working Knowledge* [online database]. Available online at http://workingknowledge.hbs.edu/pubitem.jhtml?id=3582&t=technology.

Von Hippel, E. (1997 [1988]). *The Sources of Innovation,* rev. ed. New York: Oxford University Press.

Wasserman, S., and Faust, K. (1994). *Social Network Analysis: Methods and Applications.* Cambridge: Cambridge University Press.

Watts, A. (1977). *Tao: The Watercourse Way.* New York: Random House.

Wenger, E., McDermott, R., and Snyder, W. (2002). *Cultivating Communities of Practice.* Boston: Harvard Business School Press.

Williams, L., and Cockburn, A. (June 2003). "Agile Software Development: It's about Feedback and Change." *IEEE Computer* 36:6, 39–43.

ADDITIONAL RESOURCES

Allen, T. (1977). *Managing the Flow of Technology.* Cambridge, Mass.: MIT Press.

Barabasi, A. (2002). *Linked: The New Science of Networks.* Cambridge, Mass.: Perseus.

Batagelj, V., and Mrvar, A. (1998). "Pajek—Program for Large Network Analysis." *Connections* 21:2, 47–57. Available online at http://vlado.fmf.uni-lj.si/pub/networks/doc/pajek.pdf.

Bonabeau, E., and Meyer, C. (May 2001). "Swarm Intelligence: A Whole New Way to Think about Business." *Harvard Business Review* 5, 107–114.

Borgatti, S., Everett, M., and Freeman, L. C. (1992). *UCINET IV, Version 1.0.* Columbia, SC: Analytic Technologies.

Borgatti, S. P., and Everett, M. G. (1999). "Models of Core/Periphery Structures." *Social Networks* 21: 375–395.

Burt, R. S. (1992). *Structural Holes.* Cambridge, Mass.: Harvard University Press.

Cailliau, R., and the World Wide Web Consortium (W3C). (1995). "A Little History of the World Wide Web from 1945 to 1995." HTML timeline. Available online at http://www.w3.org/History.html.

Chandler, A. D. (1977). *The Visible Hand.* Cambridge, Mass.: Harvard University Press.

Chesbrough, H. W. (July 2003). "A Better Way to Innovate." *Harvard Business Review* 81:7, 12–13.

Cohen, D., and Prusak, L. (2001). *In Good Company. How Social Capital Makes Organizations Work.* Boston: Harvard Business School Press.

Cross, R., Nohria, N., and Parker, A. (Spring 2002). "Six Myths about Informal Networks—and How to Overcome Them." *MIT Sloan Management Review* 43:3, 67–75.

Cross, R., and Prusak, L. (June 2002). "The People Who Make Organizations Go—or Stop." *Harvard Business Review* 80:6, 105–112.

Cusumano, M. A. (1998). *Microsoft Secrets: How the World's Most Powerful Software Company Creates Technology, Shapes Markets, and Manages People.* New York: Simon & Schuster.

Davenport, T., and Prusak, L. (2000). *Working Knowledge.* Boston: Harvard Business School Press.

Deloitte Consulting. (2001). *Collaborative Knowledge Networks: Driving Workforce Performance through Web-Enabled Communities.* Available online at http://www.dc.com/obx/pages.php?name=AllResearch_sub_ckn.

Dodds, P. E., Muhamad, R., and Watts, D. J. (August 2003). "An Experimental Study of Search in Global Social Networks." *Science* 301:8, 827–829.

Ebel, H., Mielsch, L., and Bornholdt, S. (February 12, 2002). "Scale-Free Topology of E-Mail Networks." *Physical Review E* 66, 035103R.

Ellickson, R. C. (1991). *Order without Law.* Cambridge, Mass.: Harvard University Press.

Farhoomand, A. F., Ng, P. S. P., and Conley, W. L. (April 2003). "Building a Successful E-Business: The FedEx Story." *Communications of the ACM* 46:4, 84–89.

Freeman, L. C., and Freeman, S. C. (1980). "A Semi-Visible College: Structural Effects of Seven Months of EIES Participation by a Social Networks Community." In *Electronic Communication: Technology and Impacts,* M. J. MacNaughton and M. M. Henderson, eds., 77–85. AAAS Symposium 52. Washington, D.C.: American Association for the Advancement of Science.

Gasser, L., and Scacchi, W. (October 15, 2003). *Continuous Design of Free/Open Source Software.* Workshop Report and Research Agenda. Irvine, Calif., and Urbana, Ill.: UCI-UIUC Workshop on Continuous Design of Open Source Software.

Girvan, M., and Newman, M. E. J. (December 7, 2001). "Community Structure in Social and Biological Networks." *Proceedings of the National Academy of Sciences USA* 99, 7821–7826.

Gloor, P. (December 1991). "CYBERMAP—Yet Another Way of Navigating in Hyperspace." In *Proceedings of the Third Annual ACM Conference on Hypertext,* 107–121. New York: ACM Press.

Gloor, P. (1992). "AACE—Algorithm Animation for Computer Science Education." In *Proceedings of the 1992 IEEE Workshop on Visual Languages,* 25–31. Seattle, Wash.: IEEE Computer Society.

Gloor, P. (2000). *Making the E-Business Transformation: Sharing Knowledge in the E-Business Company.* London: Springer.

Gloor, P. (November 2002). "Collaborative Knowledge Networks." *eJETA: The Electronic Journal for e-Commerce Tools and Applications* 1:2. Available online at www.ejeta.org.

Gloor, P., and Dynes, S. (1998). "Cybermap—Visually Navigating the Web." *Journal of Visual Languages and Computing* 9, 319–336.

Gloor, P., Laubacher, R., Dynes, S., and Zhao, Y. (2003). "Visualization of Interaction Patterns in Collaborative Knowledge Networks for Medical Applications." In *Proceedings of the Tenth International Conference on Computer-Human Interaction,* vol. 3: *Human-Centered Computing: Cognitive, Social, and Ergonomic Aspects,* 981–985.

Gloor, P., Laubacher, R., Dynes, S., and Zhao, Y. (November 2003). "Visualization of Communication Patterns in Collaborative Innovation Networks: Analysis of Some W3C Working Groups." In *Proceedings of the Twelfth International Conference on Information and Knowledge Management,* 56–60. New York: ACM Press.

Gloor, P., and Uhlmann, P. (1999). "The Impact of E-Commerce on Developing Countries." Presentation at the Third Annual Conference on Business Information Systems (BIS), Poznán, Poland.

Granovetter, M. (May 1978). "The Strength of Weak Ties." *American Journal of Sociology* 78:6, 1360–1380.

Greenspun, P. (1999). "Using the ArsDigita Community System." *ArsDigita Systems Journal* [online only]. Available online at http://eveander.com/arsdigita/asj/using-the-acs.

Greve, A. (2004). "Creativity in Social Networks: Combining Knowledge in Innovations." In *Creativity and Problem-Solving in the Context of Business Management. A Festschrift in Honor of Geir Kaufmann for His 60-Year Anniversary,* 143–163. Bergen, Norway: Fagbokforlaget.

Guimera, R., Danon, L., Diaz-Guilera, A., Giralt, F., and Arenas, A. (2003). "Self-Similar Community Structure in Organizations." *Physical Review E* 68, 065103R.

Hagel, J., and Armstrong, A. (1997). *Net Gain: Expanding Markets through Virtual Communities.* Boston: Harvard Business School Press.

Jarvenpaa, S. L., Knoll, K., and Leidner, D. E. (1998). "Is Anybody Out There? Antecedents of Trust in Global Virtual Teams." *Journal of Management Information Systems* 14:4, 29–64.

Kleiner, A. (2002). "Karen Stephenson's Quantum Theory of Trust." *strategy+business* 29, 02406R.

Kratzer, J., Leenders, R. T. A. J., and van Engelen, J. M. L. (2004). "Stimulating the Potential: Creativity and Performance in Innovation Teams." *Journal of Creativity and Innovation Management* 13, 63–70.

Larner, J. (1999). *Marco Polo and the Discovery of the World.* New Haven, Conn.: Yale University Press.

Lee, G. K., and Cole, R. E. (October 25, 2000). *The Linux Kernel Development as a Model of Knowledge Development.* Working Paper. Berkeley, Calif.: Haas School of Business, University of California, Berkeley.

Leenders, R. T. A. J., van Engelen, J. M. L., and Kratzer, J. (2003). "Virtuality, Communication, and New Product Team Creativity: A Social Network Perspective." *Journal of Engineering and Technology Management* 20, 69–92.

Lueg, C., and Fisher, D. 2003. *From Usenet to CoWebs, Interacting with Social Information Spaces.* London: Springer.

Malone, T. W., and Crowston, K. (1994). "The Interdisciplinary Study of Coordination." *ACM Computing Surveys* 26:1, 87–119.

Malone, T. W., Crowston, K. G., and Herman, G., eds. (2003). *Organizing Business Knowledge: The MIT Process Handbook.* Cambridge, Mass: MIT Press.

Malone, T. W., and Laubacher, R. (September—October 1998). "The Dawn of the E-Lance Economy." *Harvard Business Review* 76:5, 144–152.

Malone, T. W., Laubacher, R. J., and Scott Morton, M. S., eds. (2003). *Inventing the Organizations of the 21st Century.* Cambridge, Mass.: MIT Press.

Meyerson, D., Weick, K. E., and Kramer, R. M. (1996). "Swift Trust and Temporary Groups." In *Trust in Organizations: Frontiers of Theory and Research,* R. M. Kramer and T. R. Tyler, eds., 166–195. Thousand Oaks, Calif.: Sage.

Ogger, G. (1978). *Kauf dir inen Kaiser—die Geschichte der Fugger.* Munich, Germany: DroemerKnaur.

O'Mahony, S. (2003). "Guarding the Commons: How Community Managed Software Projects Protect Their Work." *Research Policy* 32, 1179–1198.

O'Mahony, S. (in press). "Developing Community Software in a Commodity World." Chapter 8 in *Frontiers of Capital: Ethnographic Reflections on the New Economy,* M. Fisher and G. Downey, eds. Social Science Research Council.

Orlikowski, W., and Yates, J. (1994). "Genre Repertoire: The Structuring of Communicative Practices in Organizations." *Administrative Science Quarterly* 39, 541–574.

Padgett, J., and Ansell, C. (May 1993). "Robust Action and the Rise of the Medici, 1400–1434." *American Journal of Sociology* 98, 1259–1319.

Pyysiäinen, P., Paasivaara, M., and Lassenius, C. (2003). *Coping with Social Complexity in Distributed Software Development Projects.* Helsinki, Finland: Helsinki University of Technology.

Schein, E. H., with Kampas, P. J., DeLisi, P., and Sonduck, M. (2003). *DEC Is Dead, Long Live DEC: The Lasting Legacy of Digital Equipment Corporation.* San Francisco, Calif.: Berrett-Koehler.

Seligman, A. B. (1998). "Trust and Sociability: On the Limits of Confidence and Role Expectations." *American Journal of Economics and Sociology* 57:4, 391–404.

Shepherd, G., and Shepherd, G. (June 25, 2002). "The Family in Transition: The Moral Career of a New Religious Movement." Presentation at the 2002 CESNUR Conference, Salt Lake City, Utah.

Siau, K., and Shen, Z. (April 2003). "Building Customer Trust in Mobile Commerce." *Communications of the ACM* 46:4, 91–94.

Tuckman, B. W. (1965). "Developmental Sequence in Small Groups." *Psychological Bulletin* 63, 384–99.

Tuomi, I. (2003). *Networks of Innovation.* New York: Oxford University Press.

Tyler, J., Wilkinson, D., and Huberman, B. A. (2003). "*Email as Spectroscopy: Automated Discovery of Community Structure within Organizations.*" In *Communities and Technologies*, 81–96. Deventer, the Netherlands: Kluwer. Available online from HP Laboratories at http://www.hpl.hp.com/shl/papers/email/index.html.

Van der Smagt, T. (2000). "Enhancing Virtual Teams: Social Relations vs. Communication Technology." *Industrial Management and Data Systems* 100:4, 148–156.

Varghese, G., and Allen, T. (Spring 1993). "Relational Data in Organizational Settings: An Introductory Note for Using AGNI and Netgraphs to Analyze Nodes, Relationships, Partitions and Boundaries." *Connections* 16:1–2.

Von Hippel, E. (June 2002). *Horizontal Innovation Networks—by and for Users.* Working Paper 4366–02. Cambridge, Mass.: MIT Sloan School of Management. Available online at http://web.mit.edu/evhippel/www/Publications.htm.

Von Krogh, G., Ichijo, K., and Nonaka, I. (2000). *Enabling Knowledge Creation: How to Unlock the Mystery of Tacit Knowledge and Release the Power of Innovation.* New York: Oxford University Press.

Watts, D. (1999). *Small Worlds.* Princeton, N.J.: Princeton University Press.

Watts, D. J., and Strogatz, S. H. (1998). "Collective Dynamics of 'Small-World' Networks." *Nature* 393, 440–442. Available online at http://tam.cornell.edu/SS_nature_smallworld.pdf.

Weber, M. (1947). *The Theory of Economic Organization.* New York: The Free Press.

Weber, M. (1978). *Economy and Society.* Berkeley, Calif.: University of California Press.

Weber, S. (2004). *The Success of Open Source.* Cambridge, Mass.: Harvard University Press.

Yates, J. (1989). *Control through Communication: The Rise of System in American Management.* Baltimore, Md.: Johns Hopkins University Press.

INDEX